SpringerBriefs in Applied Sciences and Technology

PoliMI SpringerBriefs

More information about this series at http://www.springer.com/series/11159
http://www.polimi.it

Lone Feifer · Marco Imperadori
Graziano Salvalai · Arianna Brambilla
Federica Brunone

Active House: Smart Nearly Zero Energy Buildings

POLITECNICO
MILANO 1863

Springer

Lone Feifer
Active House Alliance
Brussels
Belgium

Marco Imperadori
Department of ABC
Politecnico di Milano
Milan
Italy

Graziano Salvalai
Department of ABC
Politecnico di Milano
Milan
Italy

Arianna Brambilla
School of Architecture, Design and Planning
University of Sydney
Sydney, NSW
Australia

Federica Brunone
Department of ABC
Politecnico di Milano
Milan
Italy

ISSN 2191-530X ISSN 2191-5318 (electronic)
SpringerBriefs in Applied Sciences and Technology
ISSN 2282-2577 ISSN 2282-2585 (electronic)
PoliMI SpringerBriefs
ISBN 978-3-319-90813-7 ISBN 978-3-319-90814-4 (eBook)
https://doi.org/10.1007/978-3-319-90814-4

Library of Congress Control Number: 2018940653

Printed on acid-free paper

This Springer imprint is published by the registered company Springer International Publishing AG part of Springer Nature
The registered company address is: Gewerbestrasse 11, 6330 Cham, Switzerland

Preface

After some years of presence in the Active House (AH) Alliance, being the only professor representing Politecnico di Milano and also the Mediterranean climatic zones, I decided to propose this book in order to fix a milestone after previous years of activities and evolution and also to show the direct approach that my team has given to AH in the South European climatic zone.

Lone Feifer, current General Secretary of the Alliance and also a great expert about innovative approaches to sustainability in the field of constructions, has immediately accepted with enthusiasm this challenge in order to share and spread in the Academic community all the best practices that AH has clearly collected in recent years.

Therefore, the book will show what an Active House is and why we should go "beyond" a passive approach and imagine that the buildings of the future, smart and innovative, are in many cases already existing.

"People first!" are the magic words that sustain this approach where three main domains are studied at the same time: Comfort, Energy, and Environment.

I would like to tell that for me, being also a designer since almost 25 years, "sustainability" has been always a key issue and a *conditio sine qua non* of every architecture. With AH approach, we could imagine that our goal is also "Planet Earth first!" where humans are part of a delicate eco-system. Designers, clients, developers, etc., should be harmoniously part of it because the place where they live and that they transform must have *empathy* and *equilibrium* instead of the classic approach of *extraction* and *consume*.

This is clearly a different philosophical approach where buildings, new or renovated, can be balanced with natural, renewable energies and become "concentrators-distributors" of energies instead of being aggressive consumers of ground and resources.

Will it work?

As a matter of fact, it is already working and the shown case studies—some of them on the market and some of them experimental—give a clear answer, different by climatic zone: different approaches from the *status quo* are possible, and a turning point, a change, has been set and the alliance, among designers, institutions, industries, academies, is bringing new fresh air to the world of construction.

Architecture and aesthetic itself are very much sustained in AH where the environmental constraints are not penalizing the forms but the morphology and shapes are the result of a continuous and iterative process, which has a central role to give occupants comfort and respect natural resources.

So, reducing houses' impact on the Planet can be achieved without reducing our comfort.

Let's see how it works, be Active!

Milan, Italy Marco Imperadori

Acknowledgements

This book is the result of an intense cooperation among researchers, architects, engineers, designers, and experts from Active House Alliance and Politecnico di Milano. We are glad and proud of this outcome, whose purpose was to witness years of activities and high-profile design practices, inspired by a holist approach for high energy efficiency, sustainability, and human well-being.

Our thanks go to Active House Alliance, its founding partners and its Advisory board: Grundfos, Verband Fenster + Fassade, Danish Technological University, VELUX Group, and Great Gulf Homes; Aalborg University, Reinberg Architects, and Cenergia. In particular, we especially appreciate the collaboration of designers and industrial partners for the data sharing of the selected case history.

In the end, we would like to acknowledge all the graduating students and trainees that have been involved in the analysis, development, and construction of the Active House prototypes of Politecnico di Milano.

Contents

Chapter 1
What Is an Active House? A Vision Beyond 2020

Abstract International standards and constantly smarter practices are everyday more addressing the issue of a sustainable development. However, this virtuous approach has to involve the final users, in a newer and wider perspective: People First. This is the aim of the Active House Alliance, proposing a new vision for the future of the construction sector. Indoor comfort and occupants' wellbeing, energy efficiency and renewable sources exploitation, eco-friendly approaches are integrated into new guidelines and technical prescriptions, to help designers in developing the active buildings of the future.

Keywords Active house · People first · Guideline · AH specification
AH radar tool · Comfort · Energy · Environment

1.1 The Need for a New Perspective

What do Winston Churchill and Immanuel Kant have in common? They both offer us guidance for a new perspective on user-centric building design: *We shape our buildings; thereafter they shape us* said Churchill, and Kant's famous quote about reasoning: *All our knowledge begins with the senses, proceeds then to the understanding, and ends with reason* reminds us that buildings must be conceived as user centric. The question is: do we today design and evaluate buildings by how they function including users? Planners rarely get a chance to go back to a building after hand-over and learn from the users if their ideas worked in practice. Clients prescribe building programmes, yet how much knowledge do they possess about the building in operation?

The value of a building emerges through the user interaction and must be valued accordingly. The valuation should not primarily be by how it is taken apart after its service life, not primarily by how little energy is lost during the winter months, but primarily by its ability to support user health and wellbeing. Sensibility matters all year round—summer, spring, autumn, day and night.

This chapter is authored by Lone Feifer and Marco Imperadori.

L. Feifer et al., *Active House: Smart Nearly Zero Energy Buildings*,
PoliMI SpringerBriefs, https://doi.org/10.1007/978-3-319-90814-4_1

Can we make sense of sensibility without turning it into rocket science? Are we able to monitor buildings in use, including the user component and flexible demands without an engineering exam and a costly investment?

Indeed, we can; tools, competences, approaches and principles are here today. We can programme daylight, thermal comfort, air quality and acoustics without compromising the energy demand and environmental footprint. The key is to make these parameters tangible and thus measurable.

The Active House approach to buildings is to create healthier and more comfortable lives for the occupants, while minimising the climate impact. The design DNA is recognizable in the three principles: Energy—Environment—Comfort. The first Active House ideas emerged in 2007, followed by the first round table, held in Copenhagen in 2009. Since then, 75 Active Houses around the globe have been tested in use, researched, and documented in more than 50 peer reviewed scientific papers. The combined results and feedback from users clearly prove that we today have the products, the technology and the competences needed to build energy efficient buildings that positively affect our personal health and wellbeing.

1.1.1 People First

In 2017, the Nobel Prize in Medicine went to three researchers who have done ground-breaking research on the circadian rhythm, the biological clock for living organisms, including humans. Previously well-known from jet lag situations, these scientific revelations just after the millennial turn have dramatically affected our understanding of human circadian rhythms and our approach to tangible attributes of buildings, e.g. design for daylight. The daylight passing through the eye is 90% used for vision, and 10% affects the ganglion cells and has a direct impact on our health and wellbeing. This ground-breaking research will define a new point of departure for how we think about, plan, use and in the end, value buildings.

We are today the indoor generation, who have only recently recognized the need to design buildings for people. It is crucial to stay curious and to apply an innovation-based mindset, where you learn from what did not work as expected. We must leave behind the prejudice that users are disturbing the engineering order and forecast of a building's performance, by making people the first priority. Put people first.

1.1.2 Active Beyond Passive

There has been an intensive focus on the energy dimension of buildings for years on end, perhaps since the energy-in-buildings awareness was brought to life in the early 1970s, with the oil crisis having brought about regulatory demands concerning energy for heating. The technical and calculative approach to building to address the issue of heating efficiency was pioneered by the forward-thinking PassiveHaus pioneers. From the epicenter in the "LänderDreiEck" in the boundary area of Germany, Switzerland and Austria, they established an ideology about how good design with clever engineering tools, in combination with competent craftsmanship, could address the challenge of excessive heating demand in buildings. For decades the techniques and tools have been developed further and tested and evaluated in practice.

Nowadays, the need for visions focus always more on users rather than just on energy waste and its economy value is increasing. This is the core of Active House vision.

The Active House builds upon the knowledge and competences in the PassivHaus principles, making the next development step towards the user-centric focus. Still maximizing the use of passive technologies, such as natural ventilation, solar gains and quality craftsmanship, the Active House principle Comfort adds the sensory aspects of Daylight, Thermal Comfort and Air Quality. The Acoustic aspect is in the specifications, as well, as it is a very important aspect for completing the sensory approach of seeing, feeling, breathing and hearing.

Most importantly, the all seasons approach is introduced. Several regulatory approaches in Europe demand only specifications (buildings and components) for the winter season. However, on a cold winter day, a warm jacket is a good friend, but the jacket (a.k.a. the climate envelope) must be dynamic and flexible so that it can be adapted to suit the user's needs over a whole year and be adjusted to different users' tastes and desired temperatures.

The Active House concept places its gravity point at a combination of the human sensory, all seasons, and flexibility approaches. We must not fall for the temptation to make a one-eyed focus, on the altar of simplification. The professional competences, industrial products, and scientific proof are here today. We know that it is possible to build healthy buildings without compromising the energy demand or environmental footprint. We have moved on, and must keep looking ahead and develop further, building on the experiences of the past, while aiming at visions for the future.

We stand on the shoulders of giants, rather than behind them, so they do not block our line of vision to a holistic approach to sustainable building design.

1.2 The Alliance—A Global Conversation

In 2007 the first concept of how we could apply a holistic point of departure to the design of sustainable buildings was launched. In 2009, the first round table was held in Copenhagen, and the pioneer discussions mounted into the forming of a member-based alliance named the Active House Alliance in 2011.

The founding partners were: Grundfos, Verband Fenster+Fassade, Danish Technological University, VELUX Group, and Great Gulf Homes.

The first Advisory board: Aalborg University, Reinberg Architects, Great Gulf Homes, Grundfos, VELUX Group, Cenergia.

The first Board of directors: Hunter Douglas, 3h Architecture, VELUX Group, Verband Facade+Fenster, Danish Technological University, Jeld-Wen, FEMIB, Rockwool International.

Today the Alliance is a global partnership of more than 40 knowledge institutions, designers, engineers and industries and developers. The members have tested the Active House principles and specifications in full scale demonstration projects, more than 30 in 12 countries from 2009 to 17, and established an Active House label for the broad market on housing and smaller buildings (Fig. 1.1). Its borders are under an ongoing expansion, beyond Europe and towards other countries: two samples of Active House have been already validated in Canada (Chap. 5); in China, the Alliance's activities among industries and construction experts have started, and are giving the first results (Fig. 1.2), while the first dialogues are under development with the Australian institutions and stakeholders.

There have been a number of scientific studies into the Active House standard, including such areas as daylight design, the sociology of indoor comfort, and organizational partnerships for user-centered design. For example, Lara Anne Hale's Ph.D. dissertation entitled "Experimental Standards in Sustainability Transitions: Insights from the Building Sector" expounds her research on Active House in three articles, addressing: (1) the legitimation of comfort specifications in the building industry and among policy makers; (2) the awareness of sustainable design value brought to those who live in or work with Active Houses (Hale 2018); and (3) the need for user-centered design of technologies in smart buildings. According to Dr. Hale.

> Active House is the next frontier in the built environment. Rising to challenges set forth in the United Nations Sustainable Development Goals, Active House presents sustainable building solutions that balance energy, environment and comfort, while centering design on the building users, us human beings. We as people have sustainability concerns, but we also deserve products and services that have taken our health and well-being into consideration. Such design innovations help us to orient our future visions to include social sustainability and interactive design and inspire people to take part in the shaping of their own future.

Today there are 75 Active Houses in 19 countries; 31 with radar, tested in use and 14 labelled projects (Fig. 1.3). The label was introduced in 2016. Verifiers of label passed diplomas in 2017.

There are seven national alliances, doubling the member crowd into a global community of partners who aim to scale sustainable cities—as per United Nations (UN) Sustainable Development Goal (SDG) 11 to make cities and human settlements inclusive, safe, resilient and sustainable, and via the SDG 17 for strengthening global partnerships for sustainable development (Fig. 1.4). Each member recognizes that together we can do more than we can do alone and achieve what otherwise can seem impossible, including:

- Multi-stakeholder partnerships that mobilize and share knowledge, expertise, technology and financial resources, to support the achievement of the sustainable development goals in all countries, in particular developing countries;
- Encouragement and promotion of effective public, public–private and civil society partnerships, building on the experience and re-sourcing strategies of partnerships.

Fig. 1.1 The Green Solution House is an example of Active House labelled projects; its design integrates Active House principles and applies Active House specifications, as shown by the Active House Radar of the project (Ph. Adam Mørk, AH Radar © Adam Mørk)

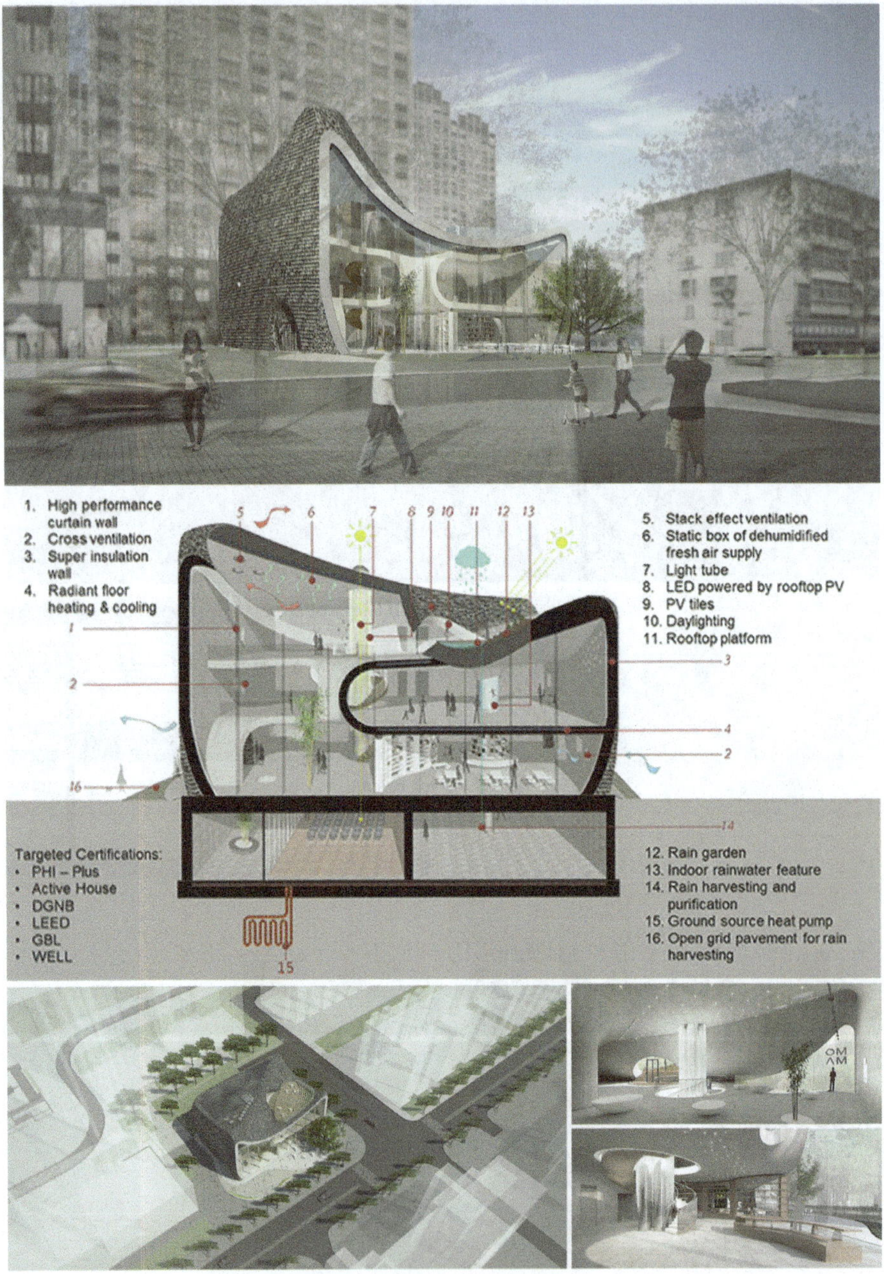

Fig. 1.2 The MOMA Building & Arts Museum, in Beijing, China. The project is the latest development of MOMA, designed by Zhuang Weimin. The building has been conceived in order to integrate building arts and technologies, Nature and buildings, structure and equipment, towards the achievement of a low-energy, comfortable and healthy building (© Active House Alliance)

Fig. 1.3 The Active House world map (© Active House Alliance)

Fig. 1.4 The 17 goals of United Nations for a sustainable development are integrated in the Active House vision, for a global conversation that gathers several national alliances into a global community (Reproduced from http://www.un.org/sustainabledevelopment/sustainable-development-goals/)

1.3 AH Technical Specifications (2013 Edition)

Today, the world is facing many environmental challenges. Natural resources are scarce, global warming needs to be tackled, and well-known sources of energy are being depleted. Meanwhile, there is a need to meet essential human needs for a healthy, comfortable indoor climate. Active House seeks to respond to these needs, through innovative technical approaches whilst introducing goals of architectural quality and environmental design—at the same time as providing energy efficiency.

The AH technical specifications represent the next generation in sustainable buildings, with a focus on user well-being, and outline the criteria to design an Active

House as a residential building that combines energy efficiency with specific attention to user health and comfort, indoor climate and the environment.

This version of the AH technical specifications has been updated in 2013, for its second edition. During the editing of this book, the AH technical specifications are under improvements, and a newer edition will be published by the end of the year.

1.3.1 Active House Vision

Buildings That Give More than They Take

Active House is a vision for buildings that create healthier and more comfortable lives for their occupants without impacting negatively on the climate—moving us towards a cleaner, healthier and safer world.

The Active House vision defines highly ambitious long-term goals for the future building stock. The purpose of the vision is to unite interested parties based on a balanced and holistic approach to building design and performance, and to facilitate cooperation on such activities as building projects, product development, research initiatives and performance targets that can move us further towards the vision.

The Active House principles propose a target framework for how to design and renovate buildings that contribute positively to human health and well-being by focusing on the indoor and outdoor environment and the use of renewable energy. An Active House is evaluated based on the interaction between energy consumption, indoor climate conditions and impact on the environment.

Comfort—creates a healthier and more comfortable life. An Active House creates healthier and more comfortable indoor conditions for the occupants, ensuring a generous supply of daylight and fresh air. Materials used have a neutral impact on comfort and indoor climate.

Energy—contributes positively to the energy balance of the building. An Active House is energy efficient. All energy needed is supplied by renewable energy sources integrated in the building or from the nearby collective energy system and electricity grid.

Environment—has a positive impact on the environment. An Active House interacts positively with the environment through an optimized relationship with the local context, focused use of resources, and its overall environmental impact throughout its life cycle (Fig. 1.5).

1.3.2 Key Principles of Active House

Buildings with Unique Combination

An Active House will always be a unique combination of the three principles: Comfort, Energy and Environment. It is precisely the combination, or 'integration', of

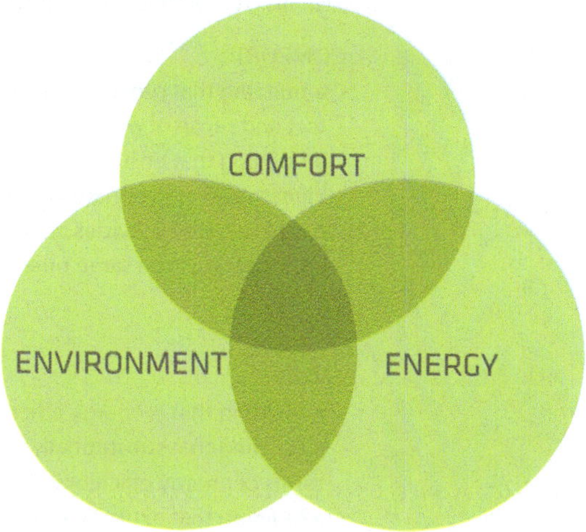

Fig. 1.5 Active House key principle: Comfort, Energy and Environment, well integrated in the AH Vision (© Active House Alliance)

these three factors that can tell the exact story of the building's architectural quality, energy efficiency, human health, comfort and well-being, and environmental benefits. This unique integration demonstrates the ambition of the Active House.

In an Active House, the integration should provide further value to (Fig. 1.6):

- the integration of the demands of comfort, energy, environment and ecology into an attractive whole;
- the architectural quality and human well-being;
- the interactive systems and spaces adding to human enjoyment and supporting environmentally responsive family life.

1.3.3 The Radar

An Active House is the result of efforts to actively integrate the three main principles of Comfort, Energy and Environment in the design of a building and in the finished building.

The Active House Radar is a tool to show the level of ambition of each of the three main Active House principles. The integration of each principle describes the level of ambition of how much 'active' the building has become. For a building to be considered as an Active House, the level of ambition can be quantified into four levels where 1 is the highest level and 4 is the lowest.

The ambitious requirement for Active House includes all nine parameters and recommends the lowest level for each of them. As long as the parameters are better or equal to the lowest level of ambition, it is an Active House within the specific parameter.

THE ACTIVE HOUSE KEY PRINCIPLES ARE AS FOLLOWS:

COMFORT

- a building that provides an indoor climate that promotes health, comfort and sense of well-being
- a building that ensures good indoor air quality, adequate thermal climate and appropriate visual and acoustical comfort
- a building that provides an indoor climate that is easy for occupants to control and at the same time encourages responsible environmental behaviour.

ENERGY

- a building that is energy efficient and easy to operate
- a building that substantially exceeds the statutory minimum in terms of energy efficiency
- a building that exploits a variety of energy sources integrated in the overall design.

ENVIRONMENT

- a building that exerts the minimum impact on environmental and cultural resources
- a building that avoids ecological damage
- a building that is constructed of materials with focus on re-use.

Fig. 1.6 The Active House key principles (© Active House Alliance)

The AH Radar shows how all parameters and goals within each principle are dependent on and balanced against each other (Fig. 1.7).

The AH Radar is a good tool for displaying the ambition reached with the building and the calculated values (Fig. 1.8). When the building is inhabited, the Radar can also be a useful tool for monitoring, evaluating and improving the building. As a communication tool, it can provide clarity as to why the integration of parameters is important for creating Active Houses (Figs. 1.9, 1.10, 1.11).

1.3.4 Comfort

An Active House is a building that lets in abundant daylight and fresh air, thereby improving the quality of the indoor climate. This also means that the thermal environment is of high quality.

We spend 90% of our time indoors; therefore, the quality of the indoor climate has a considerable impact on our health and comfort. A good indoor climate is a key quality of an Active House. It must be an integrated part of the house design

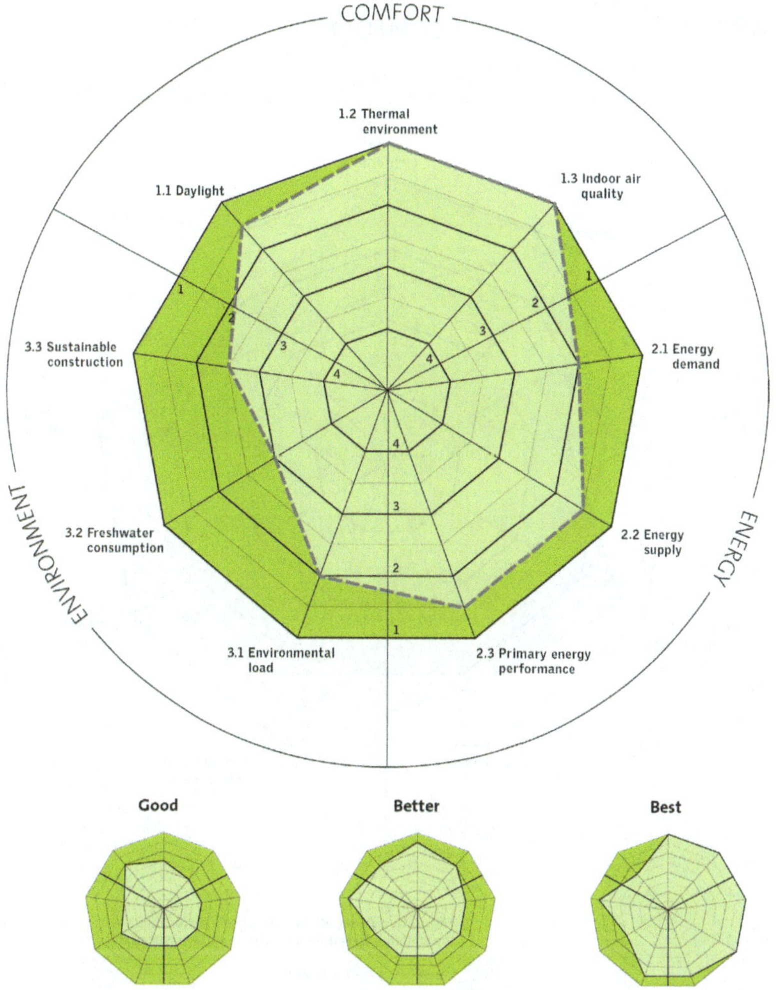

Fig. 1.7 The Active House Radar: the 4 levels of evaluation displayed in a good, better and best examples of AH Radar (© Active House Alliance)

to ensure good daylight conditions, thermal environment and indoor air quality. To support this process, the parameters in the specifications must be considered.

In order to evaluate each building's indoor climate, we utilize the four levels of ambition mentioned under Active House Radar earlier. Architects and engineers can use these levels to work towards creating their own specific levels for a building (Fig. 1.12).

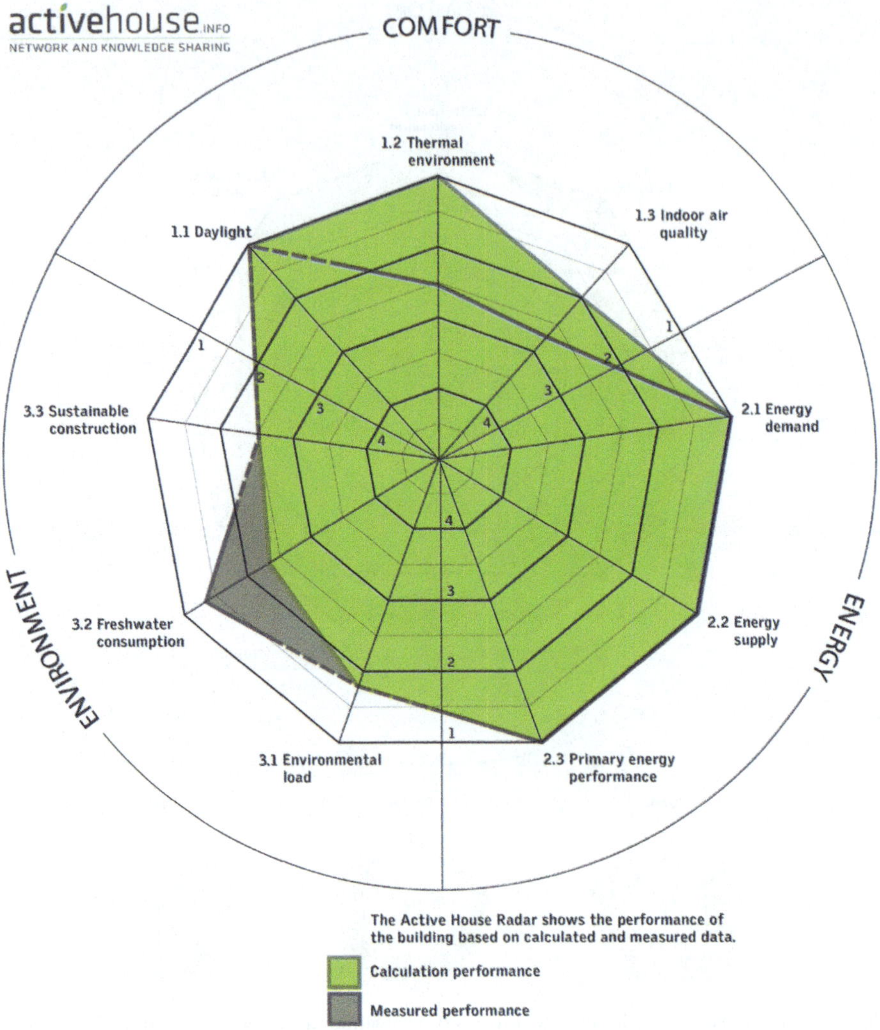

Fig. 1.8 AH Radar of Solar Decathlon project: the graph shows the differences between the calculated values during the design processes and the real measured values, reached by the building after its construction (© Active House Alliance)

Daylight

Adequate lighting and especially well-designed daylight penetration provide an array of health benefits to people in buildings. High levels of daylight and an optimized view out positively influence people's mood and well-being.

In an Active House it is thus important that the building allows for optimal daylight and attractive views to the outside. Electric lighting during daytime should rarely be necessary, which should make it possible to reduce the overall energy consumption for lighting.

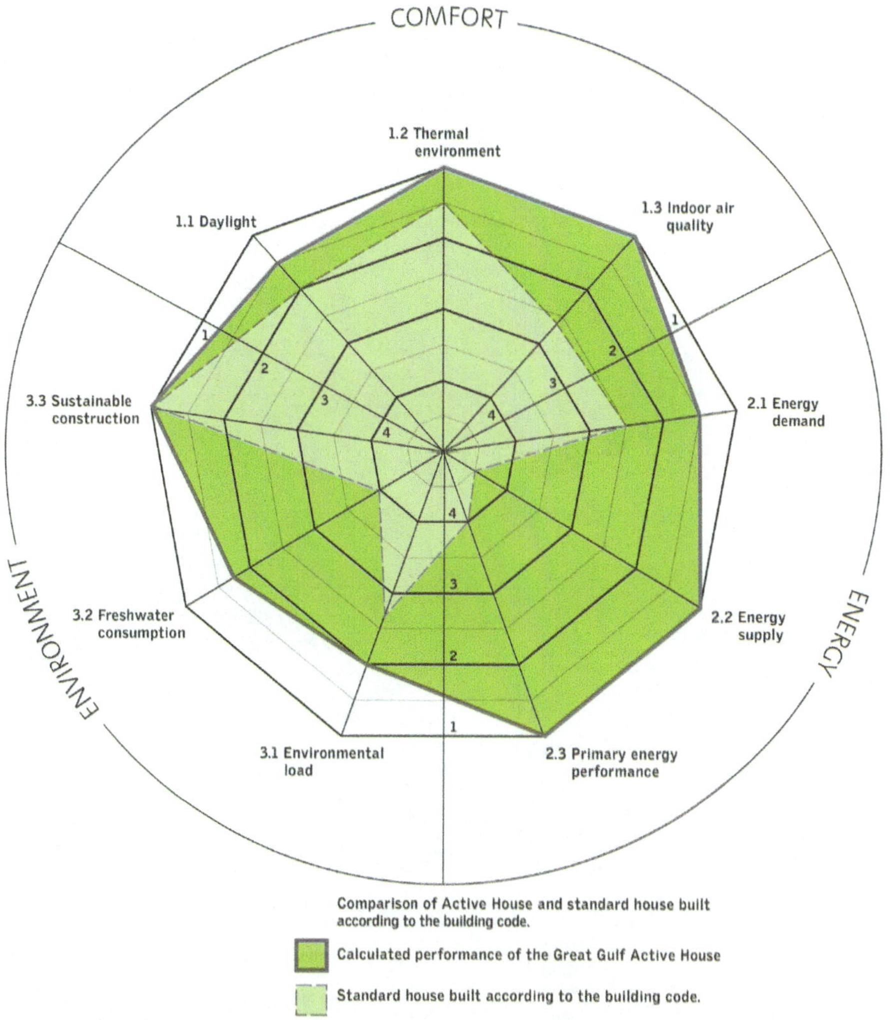

Fig. 1.9 AH Radar of Great Gulf Centennial Park project: the graph shows the Active House addition to improve the standard requirements of a common house (© Active House Alliance)

The evaluation method considers:

- The amount of daylight in a room is evaluated using average daylight factor levels on a horizontal work plane (table height approx. 0.8 m). Daylight factors are calculated with a validated daylight simulation program.
- The daylight factor is assessed room by room. Each factor is weighted to give an average daylight factor for each room. The calculation should also take neighbouring buildings into account.

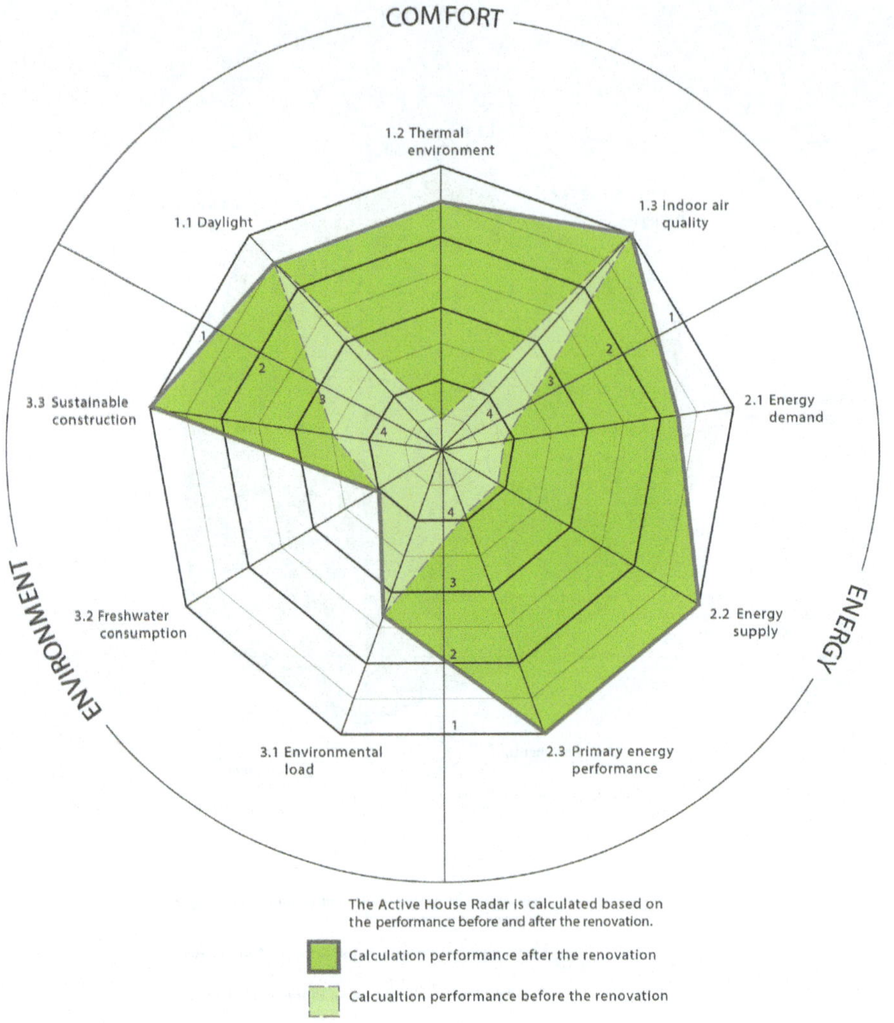

Fig. 1.10 AH Radar of The Poorters Van Montfort project: the compares a standard renovation to an Active House-based renovation (© Active House Alliance)

- The evaluation includes the living and activity zones (such as living room, work space, dining room, kitchen, bedroom or children's room).
- The room with the lowest daylight factor score sets the overall daylight factor for the building (Table 1.1).

Fig. 1.11 AH Radar of Green Solution House project: the highlights the different evaluation of the building before and after the renovation, whose strategies had followed the Active House principles (© Active House Alliance)

Thermal Environment

A pleasant thermal environment is essential for a comfortable home. Adequate thermal comfort, both in summer and winter, enhances the mood, increases performance and, in some cases (e.g. in houses for the elderly), prevents and alleviates diseases.

Active Houses should minimise overheating in summer and optimise indoor temperatures in winter without unnecessary energy use. Where possible, use simple, energy-efficient and easily maintained solutions.

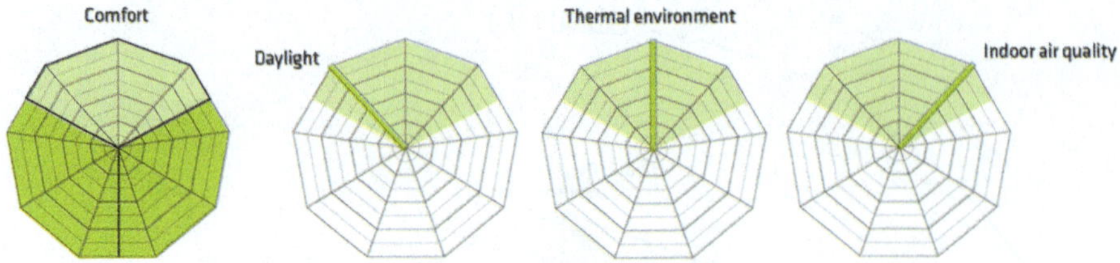

Fig. 1.12 The Active House Radar and Comfort indicators (© Active House Alliance)

Table 1.1 Daylight quantitative criteria and Active House evaluation on Daylight factor and Direct sunlight availability (© Active House Alliance)

Area	Parameter	Criteria
Comfort	Daylight factor	The amount of daylight in a room is evaluated through average daylight factor levels on a horizontal work plane: 1. DF>5% on average 2. DF>3% on average 3. DF>2% on average 4. DF>1% on average Daylight factors are calculated using a validated daylight simulation program
	Direct sun availability	For minimum one of the main habitable rooms, sunlight provision should be available between autumn and spring equinox: 1. At least 10% of probable sunlight hours 2. At least 7.5% of probable sunlight hours 3. At least 5% of probable sunlight hours 4. At least 2.5% of probable sunlight hours The evaluation is made according to British Standard BS 8206-2:2008 "Lighting for buildings—part 2: code of practice for daylight"

There are no requirements for maximum temperature in the winter (heating period) and minimum temperature in the summer (cooling period), as these are related to the behaviour of the users of the building. This is a deviation from EN15251.

The evaluation method considers:

- To objectify the risk of overheating, a dynamic thermal simulation tool is used to determine hourly values of indoor operative temperature at room level (e.g. in living rooms, kitchens and bedrooms). In dwellings without mechanical cooling systems (like central air conditioning), adaptive temperature limits are used in the summer months. This means that the maximum allowable temperature inside is linked to the weather outside: limits go up during warmer periods.
- Requirements should be met for a minimum of 95% of occupied time.
- The room with the lowest score determines the overall score for the individual parameter (Table 1.2).

Table 1.2 Thermal environment quantitative criteria and Active House evaluation on maximum operative temperature and minimum operative temperature (© Active House Alliance)

Area	Parameter	Criteria
Comfort	Maximum operative temperature	The maximum indoor temperature limits apply in periods with an outside Trm of 12 °C or more. For living rooms, kitchens, study rooms, bedrooms etc. in dwellings without mechanical air conditioning and with adequate opportunities for natural (cross or stack) ventilation, the maximum indoor operative temperatures are: 1. Ti,o < 0.33 × Trm + 20.8 °C 2. Ti,o < 0.33 × Trm + 21.8 °C 3. Ti,o < 0.33 × Trm + 22.8 °C 4. Ti,o < 0.33 × Trm + 23.8 °C Trm is the Running Mean outdoor temperature as defined in 'Sect. 3.11 External temperature, running mean of EN 15251:2007' For living rooms etc. in residential buildings with air conditioning, the maximum operative temperatures are: 1. Ti,o < 25.5 °C 2. Ti,o < 26 °C 3. Ti,o < 27 °C 4. Ti,o < 28 °C For bedrooms (especially at night time), a 2 °C lower value should preferably be used than indicated above as people are more sensitive to high temperatures when sleeping or trying to fall asleep. Also, in kitchens higher temperatures than indicated can be allowed periodically, e.g. during cooking activities. The system should be designed to achieve the values; the users can however choose other settings. Reference: EN 15251:2007
	Minimum operative temperature	The minimum indoor temperature limits apply in periods with an outside Trm of 12 °C or less. For living rooms, kitchens, study rooms, bedrooms etc. in dwellings, the minimum operative temperatures are: 1. Ti,o > 21 °C 2. Ti,o > 20 °C 3. Ti,o > 19 °C 4. Ti,o > 18 °C The system should be designed to achieve the values, the users can however choose other settings

Indoor Air Quality

Good indoor air quality can prevent humans from getting mucous membrane irritation, asthma and allergy. It can also contribute to prevent some cardiovascular diseases. High indoor air quality helps to avoid odour problems, which can positively affect the overall well-being of the building's occupants.

Table 1.3 Indoor air quality quantitative criteria and Active House evaluation of Standard fresh air supply (© Active House Alliance)

Area	Parameter	Criteria
Comfort	Standard fresh air supply	The fresh air supply shall be established according to the below limit values for indoor CO_2 concentration in living rooms, bedrooms, study rooms and other rooms with people as the dominant source and that are occupied for prolonged periods: 1. 500 ppm above outdoor CO_2 concentration 2. 750 ppm above outdoor CO_2 concentration 3. 1000 ppm above outdoor CO_2 concentration 4. 1200 ppm above outdoor CO_2 concentration

Active Houses should provide good air quality for the occupants while minimizing energy use e.g. for ventilation. This means that natural ventilation should be used where possible, or so-called hybrid systems (combination of natural and mechanical ventilation) as these systems provide the best energy performance.

Humidity only has a small effect on thermal sensation and perceived air quality in the rooms of sedentary occupancy; however, long-term high humidity indoors will cause microbial growth.

Active Houses should provide good indoor humidity levels for occupied spaces and set maximum requirements for indoor humidity. To avoid problems related to dampness and mould, it shall be guaranteed that there is sufficient extraction in rooms with periodic damp-production peaks (especially kitchens, bathrooms and toilets). The minimum exhaust air flow in these 'wet rooms' should be achievable as specified in national building codes or guidelines and the exhaust systems shall secure that the daily limit value for relative humidity in wet rooms such as bathrooms is below 80%.

The set requirements concern:

- Fresh air supply can be evaluated by examining indoor CO_2 concentrations at room level during occupancy. CO_2 is a good indicator of the amount of bio-effluents, pollutants from humans, in the air.
- Hourly values of CO_2 concentrations should be determined with a dynamic simulation tool, using standard occupancy rates (e.g. two persons in a master bedroom) and standard CO_2 production per person.
- The requirements should be met for a minimum of 95% of occupied time.
- The classification of the air quality is determined as the use-time-weighted hourly average of all room scores.
- The minimum requirements as specified in national codes should always be followed (Table 1.3).

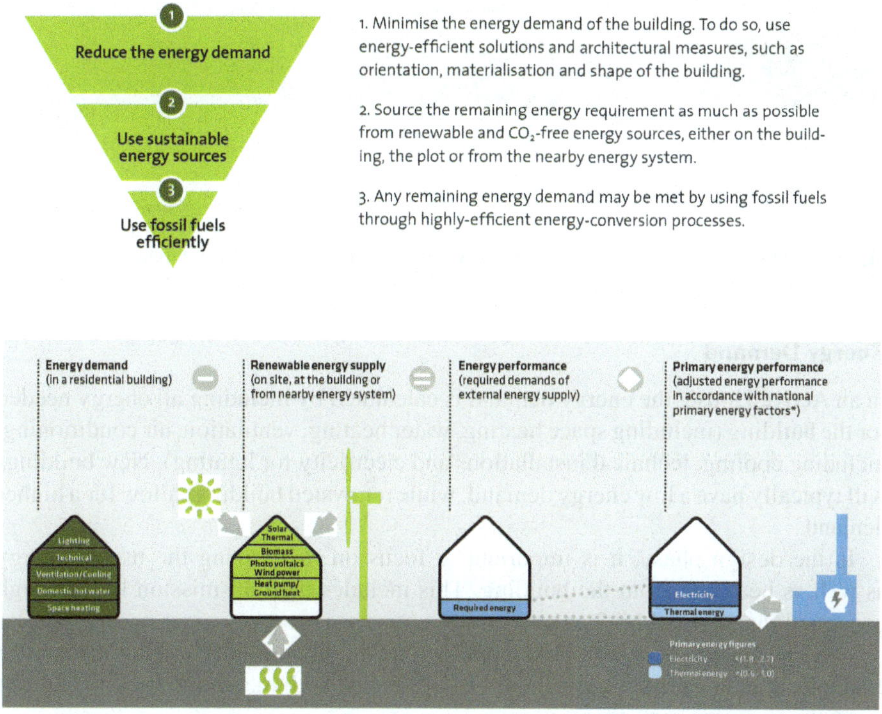

Fig. 1.13 The so-called Trias Energetica approach is the basement of the Active House Radar area of Energy: (1) Reduce the energy demand; (2) Use sustainable energy sources; (3) Use fossil fuels efficiently (© Active House Alliance)

1.3.5 Energy

An Active House is energy efficient and supplied by renewable energy sources integrated in the building or from the nearby collective energy system and electricity grid.

Globally, heating, cooling and electricity in buildings account for 40% of all energy consumption. Considering the total energy consumption throughout the whole life cycle of a building, the energy performance and energy supply are important issues in the concern about climate changes, reliability of supply and reduced global energy consumption.

The design, orientation and products for an Active House are optimized to use as little energy as possible and to utilize renewable energy sources.

The design of an Active House has to be based on the Trias Energetica approach to sustainable design (Fig. 1.13). The focus of the concept is the fact that the most sustainable energy source is saved energy (Fig. 1.14).

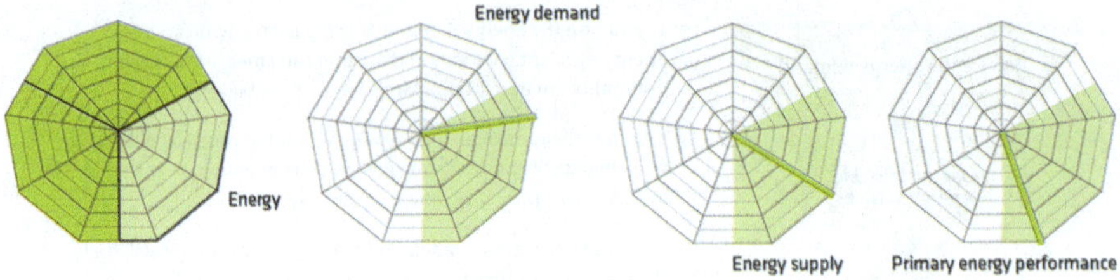

Fig. 1.14 The Active House Radar and Energy indicators (© Active House Alliance)

Energy Demand

In an Active House, the energy demand is calculated by including all energy needed for the building (including space heating, water heating, ventilation, air conditioning, including cooling, technical installations and electricity for lighting). New buildings will typically have a low energy demand, while renovated buildings allow for a higher demand.

In the design phase, it is important to focus on minimizing the use of energy as well as heat loss from the building. This includes the transmission loss through constructions, thermal bridges etc.

It is crucial to adopt a holistic approach to the use of energy. This means, for example, that an Active House should be optimized with maximum use of solutions that are not energy intensive.

Such solutions could be solar gain, daylight, natural ventilation, ventilative cooling etc. This approach is also important in regards to the need for cooling of the building. Shading of exposed facades and windows shall be established either as permanent summer shading, or as dynamic shading, such as intelligent insulation of glazed facades.

The definition of the heated floor area shall follow the national definition.

The evaluation method consists of:

- The annual energy demand includes energy demand for space heating, water heating, ventilation, air conditioning including cooling, technical installations and electricity for lighting.
- The annual energy demand shall follow the national calculation methodology. The calculation of heated floor area shall follow the national method.
- The requirements to individual products and construction elements (i.e. minimum thermal resistances, maximum thermal bridge effects and airtightness) shall at least follow requirements set in national building regulations (Table 1.4).

Energy Supply

The goal is that the energy supply to an Active House shall be based on renewable and CO_2-neutral energy sources in accordance with the energy performance classification chosen.

Table 1.4 Energy demand quantitative criteria and Active House evaluation of Annual energy demand (© Active House Alliance)

Area	Parameter	Criteria
Energy	Annual energy demand	1. $\leq$40 kWh/m^2 2. $\leq$60 kWh/m^2 3. $\leq$80 kWh/m^2 4. $\leq$120 kWh/m^2

Table 1.5 Energy supply quantitative criteria and Active House evaluation on Origin of energy supply (© Active House Alliance)

Area	Parameter	Criteria
Energy	Origin of energy supply	1. 100% or more of the energy used in the building is produced on the plot or in a nearby system 2. $\geq$75% of the energy used in the building is produced on the plot or in a nearby system 3. $\geq$50% of the energy used in the building is produced on the plot or in a nearby system 4. $\geq$25% of the energy used in the building is produced on the plot or in a nearby system

There are no specific requirements to where and how the renewable energy is produced. It must, however, be documented that the energy comes from renewable energy in the energy system.

The evaluation method consists of:

- The annual energy supply from renewable energy and CO_2-free energy sources shall be calculated and divided into the different sources (solar thermal, heat pumps, biomass, PV, wind etc.).
- The definition of renewable energy sources follows the EU Directive on the promotion of the use of energy from renewable sources (2009/28/EC of 23 April 2009).
- Requirements to performance of the individual renewable source shall follow the requirements in national building legislation. As an alternative to national requirements, the requirement in the EU Directive on the promotion of the use of energy from renewable sources (2009/28/EC of 23 April 2009) can be used.
- Renewable energy sources can either be in the building, on the site, from a nearby energy system or electricity grid and can be a mix of these sources.
- The definition of nearby system and boundaries for renewable energy on the plot follows the national or European definitions (Table 1.5).

Primary Energy Performace

The annual primary energy performance shall be based on national figures on primary energy. The calculation shall include energy demand for the building as well as the energy supply from renewable energy.

Table 1.6 Primary energy performance quantitative criteria and Active House evaluation on Origin of energy supply (© Active House Alliance)

Area	Parameter	Criteria
Energy	Annual primary energy performance	1.<0 kWh/m^2 for the building 2. 0–15 kWh/m^2 for the building 3. 15–30 kWh/m^2 for the building 4. ≥30 kWh/m^2 for the building

The evaluation method consists of:

- The primary energy performance = (energy used − renewable energy supply) × national primary energy factors.
- The energy demand and the energy produced by renewable sources must be monitored on a yearly basis. Metering devices are to be used for all types of energy production/consumption at the building level.
- The annual energy performance evaluation shall specify the supply from individual renewable energy sources integrated into the building and energy supply from local energy system and the share of renewable energy as well as the CO_2 emissions from the local energy system.
- Calculation of primary energy and CO_2-emissions shall be based on the national calculation methodology, using nationally adopted efficiency/conversion and emission factors, as well as climate data (Table 1.6).

1.3.6 Environment

An Active House will have an impact on the environment—it should be as positive as possible. This means that any harm to environment, soil, air and water should be minimized.

The challenges we face regarding the environment are a reality on a local, regional and global level. Global environmental resources are under pressure from overconsumption and pollution. The pressure is felt at a regional and a local level too.

When developing an Active House, it is important to ensure that such challenges are considered at both global, regional and local levels. This is important in order to ensure a new generation of buildings and products that aim to have a positive impact on the environment.

Consideration should be given in the design phase for how Active Houses use building materials and resources.

It is also possible to consider the local building culture and behaviour in and around the local buildings as well as traditions, climate and ecology. This is relevant when working on improving the building's exterior and interior relations to the cultural and ecological site-specific context (Fig. 1.15).

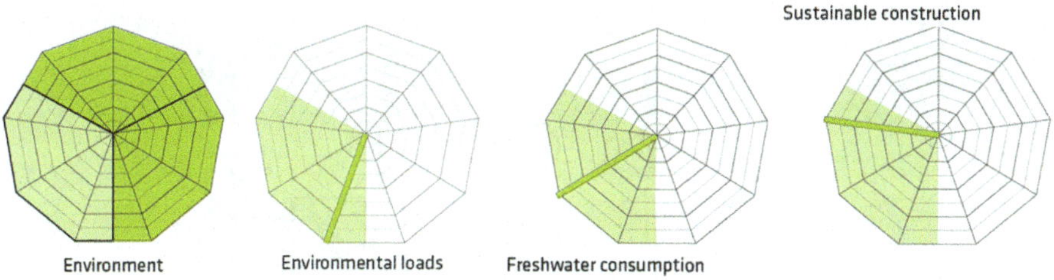

Fig. 1.15 The Active House Radar and Environmental indicators (© Active House Alliance)

The key parameters to consider within resource and emissions are:

- Consumption of non-renewable energy resources
- Environmental loads from emissions to air, soil and water
- Freshwater consumption.

As an AH needs rigorous evaluation of its performances, during this process, it is important to consider the consumption of energy resources and the emissions to air, soil and water through a Life Cycle Assessment in accordance with the EN 15643 series on sustainable construction or with ISO 14040.

The building's life cycle is considered at the following stages:

- Production of building materials
- Construction processes
- Operation of the building and maintenance of the building construction and fabric
- End of life of building materials
- Transport and site processes may be omitted.

At least, all major building components should be considered, that is

- Outside walls, roofs, slabs, foundations, windows and doors
- Inner walls, floors and ceilings
- Major technical components (heat generators etc.).

The estimated service life of the building should be in accordance with local standards. Active House suggests 50 years as benchmark. The estimated service life of all building components should be in accordance with local standards and experiences.

Environmental Product Declaration (EPD) and average data from public sources or software tools can be used as long as they are applicable to the country or region.

The following impact categories has to be evaluated:

- Resource consumption:
  - Primary energy consumption (non-renewable)
  - Primary energy consumption (renewable)

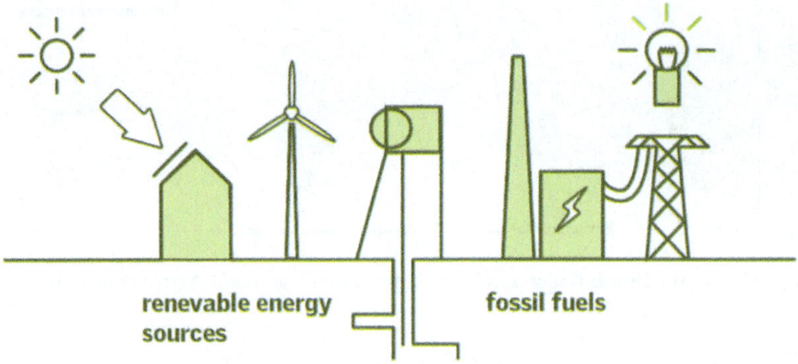

Fig. 1.16 The Active House assessment of environmental loads: Primary energy (PE) schematic design (© Active House Alliance)

- Impact categories (emissions):
  - Global warming potential (GWP)
  - Ozone depletion potential (ODP)
  - Photochemical ozone creation potential (POCP)
  - Acidification potential (AP)
  - Eutrophication potential (EP)
  - Other impact categories

Enviromental Loads

The process of constructing a new building causes various emissions to air, soil and water, which have different impacts on the environment. When constructing an Active House and conducting a Life Cycle Assessment, it is important to know and consider the different impact categories of these emissions, which may have serious environmental effects.

Primary energy (PE)

Primary energy (PE) measures the total amount of primary energy directly withdrawn from the hydrosphere, atmosphere, geosphere or energy source without any anthropogenic change, including both non-renewable and renewable resources. Primary energy is expressed in Mega Joules (MJ) and as net calorific value (Fig. 1.16).

Global warming potential (GWP)

The accumulation of so-called greenhouse gases in the troposphere causes increased reflection of infrared radiation from the earth's surface. Consequently, the temperature on the earth's surface rises. This phenomenon is referred to as the 'greenhouse effect', affecting human health, ecosystems and society in general. The global warming potential groups together gases in relation to the impact of carbon dioxide (CO_2) (Fig. 1.17).

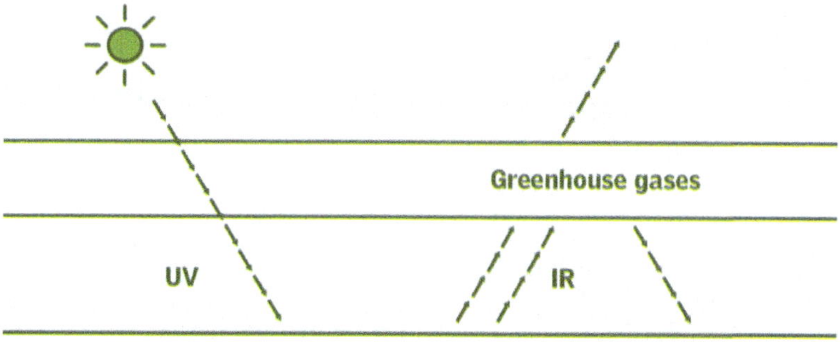

Fig. 1.17 The Active House assessment of environmental loads: Global warming potential (GWP) schematic design (© Active House Alliance)

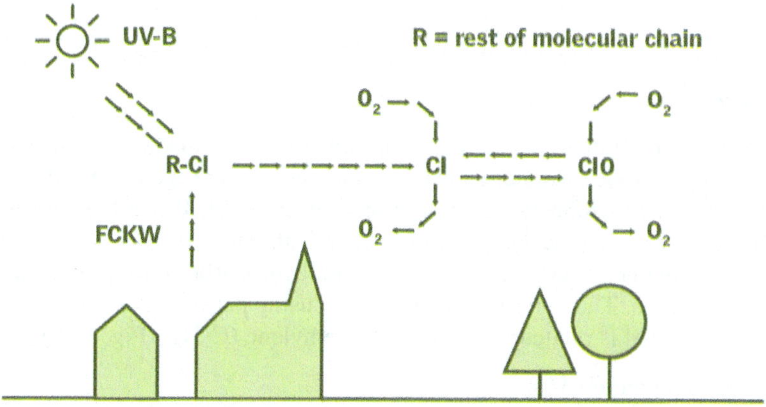

Fig. 1.18 The Active House assessment of environmental loads: Ozone depletion potential (ODP) schematic design (© Active House Alliance)

Ozone depletion potential (ODP)

Ozone (O_3) occurs as a trace gas in the stratosphere (10–50 km altitude) and absorbs solar UV radiation. However, human emissions induce the thinning of the stratospheric ozone layer since certain gases, such as halocarbons, work as catalysts degrading ozone to oxygen. Thus, the transmission of UV-B radiation is increased, with potentially harmful impacts on human health, terrestrial and aquatic ecosystems, causing for example DNA-damage, cancer (especially skin cancer) and eye irritation, crop failures and the decrease of planktons. The ozone depletion potential groups together the impact of various ozone depleting gases (Fig. 1.18).

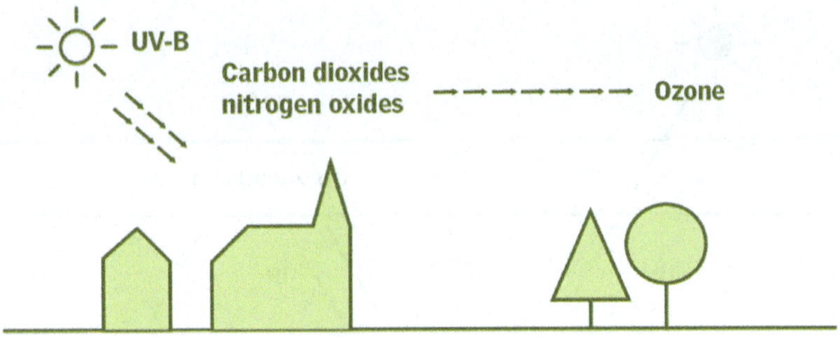

Fig. 1.19 The Active House assessment of environmental loads: Photochemical ozone creation potential (POCP) schematic design (© Active House Alliance)

The reference variable used is R11 (trichlorofluoromethane CCl3F).

Photochemical ozone creation potential (POCP)

A higher concentration of ozone in the troposphere (0–15 km altitude), the so-called summer smog, is toxic to humans and may cause damage to vegetation and materials.

When exposed to solar radiation, nitrogen oxide and hydrocarbons form round level (tropospheric) ozone in a complex chemical. This process is called photochemical oxidant creation. Nitrogen oxides and hydrocarbons are produced during partial combustion. The latter is also created by using petrol or solvents. The ozone formation potential is related to the impact of ethylene (C_2H_4) (Fig. 1.19).

Acidification potential (AP)

Acidification of soil and water results from the conversion of airborne pollutants into acids. The major acidifying pollutants are Sulphur dioxide (SO_2), nitrogen oxides (NOx) and their acids (H_2SO_4 and HNO_3). These gases are generated during combustion processes in power stations and industrial buildings, in homes, by cars and small consumers. Acidification has a wide range of impacts on vegetation, soil, groundwater, surface waters, biological organisms, ecosystems and building materials, e.g. forest decline and acid rain. The acidification potential groups together all the substances contributing to acidification in relation to the impact of Sulphur dioxide (SO_2) (Fig. 1.20).

Eutrophication (EP)

Eutrophication means excessive fertilization and describes the concentration of nutrients and nutrient enrichment in an ecosystem, which may cause an undesirable shift in species composition and elevated biomass production. The major nutrients are nitrogen (N) and phosphorus (P). These substances are contained in fertilizers, nitrogen oxides from combustion engines, domestic wastewater, industrial waste and wastewater. Plants in excessively fertilized soils exhibit weakening of their tissue and lower resistance to environmental influences. In aquatic ecosystems, increased biomass production may lead to depressed oxygen levels, because of the additional

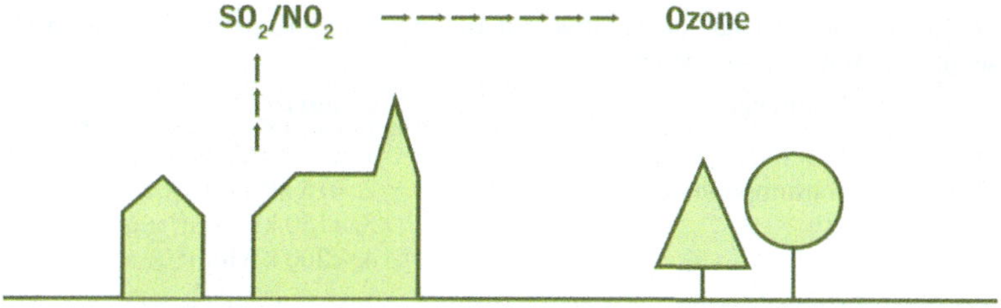

Fig. 1.20 The Active House assessment of environmental loads: Acidification potential (AP) schematic design (© Active House Alliance)

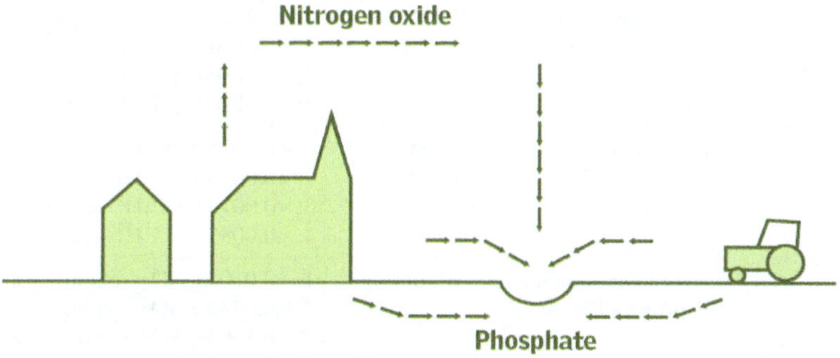

Fig. 1.21 The Active House assessment of environmental loads: Eutrophication (EP) schematic design (© Active House Alliance)

consumption of oxygen in biomass decomposition. This may result in fish death and the biological death of water. Furthermore, a high concentration of nitrate can render groundwater and surface waters unusable as drinking water since nitrate reacts and becomes nitrite, which is toxic for human beings. The eutrophication potential groups together the substances in comparison to the impact of phosphate (PO^3) (Fig. 1.21; Table 1.7).

Freshwater Consumption

The depletion and scarcity of global freshwater resources are escalating and thus it is becoming increasingly important to consider water consumption—and treatment—during the lifetime of an Active House. For example, when freshwater is saved, it results in wastewater savings as well and both issues are part of the specifications—wastewater as qualitative indications. Freshwater consumption can be reduced by installation of water-saving tabs, use of grey or rain water for toilets and gardening, and the use of easy-to-clean surfaces (Table 1.8).

Sustainable Construction

When designing an Active House, it is important to evaluate recycled content and sourcing.

Table 1.7 Environmental loads quantitative criteria and Active House evaluation on Origin of energy supply (© Active House Alliance)

Area	Parameter	Criteria
Environment	Building's primary energy consumption during entire life cycle	1. <-150 kWh/m^2 × a 2. <15 kWh/m^2 × a 3. <150 kWh/m^2 × a 4. <200 kWh/m^2 × a
	Global warming potential (GWP) during building's life cycle	1. <-30 kg CO_2-eq./m^2 × a 2. <10 kg CO_2-eq./m^2 × a 3. <40 kg CO_2-eq./m^2 × a 4. <50 kg CO_2-eq./m^2 × a
	Ozone depletion potential (ODP) during building's life cycle	1. <2.25E-07 kg R11-eq./m^2 × a 2. <5.3E-07 kg R11-eq./m^2 × a 3. <3.7E-06 kg R11-eq./m^2 × a 4. <6.7E-06 kg R11-eq./m^2 × a
	Photochemical ozone creation potential (POCP) during building's life cycle	1. <0.0025 kg C_3H_4-eq./m^2 × a 2. <0.0040 kg C_3H_4-eq./m^2 × a 3. <0.0070 kg C_3H^4-eq./m^2 × a 4. <0.0085 kg C^3H^4-eq./m^2 × a
	Acidification potential (AP) during building's life cycle	1. <0.010 kg SO_2-eq./m^2 × a 2. <0.075 kg SO_2-eq./m^2 × a 3. <0.100 kg SO_2-eq./m^2 × a 4. <0.125 kg SO_2-eq./m^2 × a
	Eutrophication potential (EP) during building's life cycle	1. <0.0040 kg PO_4-eq./m × a 2. <0.0055 kg PO_4-eq./m^2 × a 3. <0.0085 kg PO_4-eq./m^2 × a 4. <0.0105 kg PO_4-eq./m^2 × a

Table 1.8 Freshwater consumption quantitative criteria and Active House evaluation on Origin of energy supply (© Active House Alliance)

Area	Parameter	Criteria
Environment	Minimisation of freshwater consumption during building's use	Calculation is based on the national Average water consumption per building per year 1. Improvement 50% (vs. average) 2. Improvement 30% 3. Improvement 20% 4. Improvement 10% %=(National average – building consumption) × 100/National average

Table 1.9 Sustainable construction quantitative criteria and Active House evaluation on Origin of energy supply (© Active House Alliance)

Area	Parameter	Criteria
Environment	Recyclable content	By weight, the average of recycled content for all building materials (weighted by the proportion of the material in the building) could be: 1. ≥50% 2. ≥30% 3. ≥10% 4. ≥5% 80% of the weight of the building should be accounted for. (In the recycled content, we take into account internal, pre-consumer and postconsumer recycling)
	Responsible sourcing	1. 100% of the wood used is certified (FSC, PEFC) and 80% of the new material suppliers have a certified EMS 2. 80% of the wood used is certified (FSC, PEFC) and 50% of the new material suppliers have a certified EMS 3. 65% of the wood used is certified (FSC, PEFC) and 40% of the new material suppliers have a certified EMS 4. 50% of the wood used is certified (FSC, PEFC) and 25% of the new material suppliers have a certified EMS

The recycled content is evaluated by weight and shall take into consideration 80% of the weight of the building. It includes pre-consumer, internal and post-consumer recycling.

Responsible sourcing includes the requirement to use certified sourcing either directly, like PEFC and FSC for sourcing of wood, and a supplier certification EMS for other materials (Table 1.9).

1.3.7 Qualitative Parameters

The Active House qualitative parameters are listed in the following. They contain the other important issues to consider when designing an Active House. These issues are often process-oriented; some provide guidance on how to achieve the performance level described in the quantitative part, some provide guidance on how to achieve a more holistic approach (biodiversity, culture and local setting).

Comfort: Daylight

View: Are windows located to offer the best possible views to the exterior environment (sky and surroundings)?

Visual transmittance: Are windows (providing a view to the outdoors) selected to have the highest possible visible transmittance.

Glare management: Have you been considering Transmittance, Shading, Design, Reflection, Redirection, Geometry and layout of the rooms to avoid risk of glare.

Daylight in secondary rooms: Have circulation zones and bathrooms access to daylight?

Comfort: Thermal Environment

Individual control, winter: Is it possible to adjust the temperature at room level according to momentary needs, e.g. with adjustable thermostats?

Individual control, summer: Is it possible to manually influence the thermal conditions in each room, e.g. by opening windows or adjusting solar shading? In the case of mechanical cooling systems, is it possible to adjust the temperature at room level, e.g. with adjustable thermostats?

System interface: Have the climate system interfaces (e.g. wall thermostats) been selected to be as intuitive and simple as possible?

Draught: Have ventilation openings (including windows, ventilation grilles and mechanical ventilation devices) been located and detailed so that discomfort caused by draught is minimised?

Comfort: Indoor Air Quality

Individual control: Is it possible to manually influence the air change rate in the rooms (especially living room, kitchen and bedrooms), e.g. by opening windows, and if mechanical ventilation is installed, is it possible to adjust the airflow rate at three or more levels?

Dampness: To avoid problems related to dampness and mould, is it guaranteed that there is sufficient extraction in rooms with periodic damp-production peaks (esp. kitchens, bathrooms and toilets)?

Low-emitting building materials: Have indoor climate-labelled materials been used where possible?

Comfort: Noise and Acoustics

Inside system noise: mechanical services in continuous operation is below 25 db(A) in living rooms, kitchens and other main rooms, and below 20db(A) in bedrooms, study rooms and other rooms that need extra quietness?

For inside system noise, exposure to system noise (e.g. from ventilation or heating systems) is determined with a sound pressure measurement, as described in ISO 140: 1998.

Outside noise: If the building is in a noisy environment, have sufficient sound insulation measures been taken?

Acoustic privacy: Are inner walls and floor divisions designed to reduce noise transmission between rooms?

Is there at least one room (and its entrance door) that is extra sound insulated?

Energy: Energy Demand

Demand on individual products and construction elements: Have the chosen products and construction solutions been evaluated from a cost-efficient, life-cost perspective and maintenance view?

Architectural design solutions: Have architectural design solutions been used to reach a holistic approach of the building, as well as to reach a low energy demand?

Demand on individual appliances: Have the best energy-performing solutions for appliances been chosen?

Energy: Energy Supply

Design: Has integration of renewable energy been worked with as a part of the building design and typology of the building and the plot?

Origin of energy supply: Has the energy supply been evaluated from a cost perspective, and how was the decision about the origin of the energy supply made?

Energy: Primary Energy Performace

Energy use and CO_2 emissions: Has the energy demand and the use of renewable energy been optimised in order to use the most cost-efficient solutions with the lowest CO_2 emissions?

Energy: Validation on Site

Onsite control of solutions and products: Has onsite control focused on proving that the energy solutions and products meet the designed level?

Air permeability of the building: Has the air permeability of the building been tested?

Thermal bridges: Have the thermal bridges been evaluated during the construction phase?

Qualification of the controller: Has the control been established by a certified expert?

Environment: Environmental Loads

LCA of the building: Were the results of an LCA used to optimise the design?

Environment: Freshwater Consumption

Appliances: Have water saving appliances been installed?

Use of grey or rainwater: Is grey or rainwater used to reduce the water consumption in garden, toilets and for washing?

Environment: Ecological Impacts

Job site management: Have you considered optimising the management of construction waste, to limit nuisance and pollution, to minimise resource consumption and to respect workers, in order to minimise the impact of the job site?

Disassembly: Has the building been designed so that 70% by weight of the building will be able to be reused, recycled or backfilled at its end of life? This will meet and anticipate the European Directive on construction waste.

Biodiversity: Has respect for fauna, flora and the environment been considered, in order to provide nests for birds create green vegetation, minimise chemical treatment, etc.

Environment: External Context and Accessibility

Building traditions: Does the design of the building reflect a relationship with regional building traditions? E.g., were regional materials, architectural typology and artisan ship analysed and used as design parameters?

Active outdoor living: Is the design of the building adapting to the potentials and constraints of the local climate? E.g. creating private outdoor spaces with a comfortable climate and access to sunlight that encourage healthy active outdoor living.

Streets and landscapes: Does the design affect existing streets and landscapes? E.g. provision for children to play safely outside the house and supporting the public outdoor space for local behaviour, needs and tradition?

Infrastructure: Does the infrastructure support healthy, comfortable and ecological transport? E.g. connection and distance to nearest public transport for commuting, distance to schools and supermarket and the possibility of easy and safe use of bicycles.

Accessibility: Does the design take into account physically challenged people, children, elderly people or other groups with special needs?

Ecology and land use: Does the building optimise the relationship with the local ecology and land use and at the same time minimise environmental risks? E.g. maximising surface for seepage of rainwater, minimising use of land.

Climate changes: Were possible risks caused by climate changes (storms and flooding) identified and limited in the design of the building and landscape?

Building Management

When designing an Active House, it must be ensured that the building in operation works as intended and that the users are capable of behaving as intended.

In order to ensure efficient and responsible operation of the building and to assist building users in a responsible behaviour, it is important that the building and systems work as intended, that building users are aware of actual performance and that assistance is provided to promote efficient and responsible behaviour.

Management of energy:

- Training of building users in energy-efficient behaviour. Yearly performance check of building and systems (service contract)
- User guidelines on energy-efficient operation of building and technical systems
- Continuous (hourly) monitoring and display of energy use and production
- Commissioning of building, building services and renewable energy systems during the first year of operation.

Management of indoor climate:

- Training of building users in efficient behaviour. Yearly performance check of building and systems (service contract)
- User guidelines on operation of building and technical systems Continuous (hourly) monitoring and display of performance and indoor air quality

- Commissioning of building, building services, shading and ventilation systems during the first year of operation.

Management of environment:

- Training of building users in resource-efficient and responsible environmental behaviour. Yearly performance check of building and systems (service contract)
- User guidelines on optimal maintenance and water-efficient operation of building and technical systems
- Continuous monitoring and display of water use
- Commissioning of building, building services and water systems during the first year of operation.

Bibliography

Christoffersen J, Foldbjerg P (2015). Results from post-occupancy evaluation in four single-family houses. In: Healthy buildings 2015

Eriksen KE (2013) Active House Alliance—a network for development of sustainable buildings. REHVA J 03:2013

Eriksen KE, Rode C (2013) Active House specification—evaluation of comfort, energy and environment in buildings. REHVA J

Eriksen KE (2016) Active House—a tool for development of CO_2 neutral and sustainable buildings. PLEA Los Angeles 07:2016

Foldbjerg P, Asmussen TF, Feifer L, Christoffersen J (2016a) Interaction between high daylight levels and thermal comfort in five single-family houses after a full year of measurements-9 key point summary of VELUX building monitoring. In: Clima 2016

Foldbjerg P, Pauquay S, Geuens J (2016b) Renovation of a single family house in a social housing garden city in Brussels as private-public collaboration: ambitious targets for energy, indoor climate and post-occupancy monitoring. In: Clima 2016

Foldbjerg P, Arnold Andersen P, Roy N, Christoffersen J (2017) Circadian house: a vision for homes designed to be healthy and human-centric. In: AIVC 2017

Hale L (2017) Experimental standards in sustainability transitions: insights from the building sector. (Doctoral Dissertation). Doctoral School of Organisation and Management, Copenhagen Business School, Denmark

Hale LA (2018) At home with sustainability: from green default rules to sustainable consumption. Sustainability 10(1):249

Handbook on Life Cycle Assessment, Guinée. Kluwer Academic Publishers, Dordrecht, p 82

Holzer P, Foldbjerg P (2014) Control of indoor climate systems in Active Houses. In: World sustainable buildings Barcelona

http://www.un.org/sustainabledevelopment/sustainable-development-goals/

https://www.activehouse.info/

Chapter 2
A New Paradigm for Holistic Design: Active House Prototypes at Politecnico di Milano

"*Sense* experience and *necessary* demonstration"

Galileo Galilei

Abstract Active House principles represents an affordable element that could drive the design process in different stages. At Politecnico di Milano, several experimental activities have been conducted on its prototypes buildings, in order to mind the gap between the predicted, simulated, observed, and recorded building behavior. Therefore, the AH vision—with its Specifications and Guidelines—has been applied to the validation of these two innovative small buildings: VELUXlab and C-ASA. The final aim is multiple: at first, the analysis of the building features in relation to a specific climate context (the Mediterranean area); then, the information of the users, in order to educate them to the correct use of this smart objects.

Keywords Active house · Building prototypes · Design models · Model validation · Monitoring campaign · BIG DATA

2.1 Houses for the Future… Now!

Active House has proved to be a flexible strategy to design and managing buildings: beyond Passive! But also beyond 2020 (this "horizon" is almost achieved) and beyond any rigid protocol.

Sometimes this is philosophically interesting also compared to some leading EU policies about Economy where flexibility is forbidden. On the reverse, the belief that difference and flexibility are an intrinsic richness of any design strategy stays under the AH approach allowing multiple different "platform approaches" thus multiple outputs.

This chapter is authored by Marco Imperadori.

L. Feifer et al., *Active House: Smart Nearly Zero Energy Buildings*, PoliMI SpringerBriefs, https://doi.org/10.1007/978-3-319-90814-4_2

First of all, this is due to the extremely different climates that Europe has (and, in general, the World has, if we want to apply AH on a larger geographic scale): imagine Palermo in summer or Warsaw in winter. It is impossible to design a building without taking in consideration, deeply, such differences, in all the seasons.

Climate change, despite irresponsible politically interested positions, is a clear evidence, proved scientifically, and global warming is under our eyes, every day. Therefore, some strategies, which were normally more concerned about cold and winter problems (Central/North Europe), are now facing hotter summer and overheating in middle seasons, which means a crazy application of Air Conditioning and cooling in regions that with natural ventilation could easily solve that.

Maybe, in this case, North EU countries are getting "more Mediterranean" and could learn from South EU ones and recent best practices (while south EU have to adapt to increasing tropical like climate). AH approach and its adaptation to South EU climate (like Italy, Greece, Portugal, Spain…) are extremely interesting because they could be helpful to anticipate the scenario to Northern EU. So, "learning from PIGS"… This not so gentle acronym, used several times to penalize South EU countries when speaking about Economy, but South EU nations have a long tradition of master buildings, profiting from vernacular knowledge, using as much as possible the beautiful Mediterranean natural ventilation (and natural light) to provide as much as possible cooling without any fossil power.

Is it still possible today, with an upgrade and changes in building envelopes?

The answer is yes! AH strategies could be the best to match Energy, Comfort and Environment optimization at the same time in a warmer climate (induced by actual Anthropocene era) where natural sources of climatization can be used as much as possible.

The experiments followed by Politecnico di Milano in the framework of research connected with AH are different on purpose but in general, let's resume that a common strategy in Southern Europe Climate should be:

- DEFENSIVE in DEEP WINTER, well-insulated envelopes, mechanical ventilation with heat recovery, inner mass to store heat. Possible sun catching
- DEFENSIVE IN DEEP SUMMER, correct thermal delay in envelopes and thermal attenuation of heat wave, mass inside the slabs to reduce heat peaks, fixed or movable sun shadings to protect glazed façades, mechanical ventilation with energy/cooling recovery.
- NATURAL VENTILATION in MIDDLE SEASONS, and whenever outside conditions allow it.

Basic rules? Of course yes, in principle, because then any single project will be different and any single strategy will have its own variables, correct design, site-specific adaptation and so on, but these very simple rules show how the building cannot, and must not, be: "rigid", "passive" and "static". It is naturally "active"… Re-active? Interactive? Pro-active and so on, towards external and internal loads, adapting itself very quickly (and not inertially), following seasonal variations, which are clearly visible in recent years.

This will allow us to produce "cyclical" building, dynamic, able to follow the seasons' cycle but also to quickly adapt to sudden variations of normal foreseen conditions.

In this book, we want to show several cases, and share them, as experiments, which have all in common these very simple principles. Then, each one has a very different "tectonics", layering, surface, shape, and use:

- VELUXlab: Office Building, almost 100 m^2, steel structure, steel and concrete slabs, different layering for the envelope, insulation mix: polyurethane, rock wool, wood wool;
- C-ASA: Office Building, 20 m^2, CLT-wood structure, different layering for the envelope, insulation only by thermos-reflective multilayer systems;
- R-HOME[1]: Housing, wood platform frame prefab structure, multi-layered, insulations mainly in cork and wood fibers panels;
- MAISON VERTE[2]: Housing, 350 m^2, steel structure, steel and concrete slabs, multi-layer dry envelope, insulation rock wool and neopor.

The purpose of these differences is to show that there is not an exclusive technology, material, or assembly strategy, no limitation of shape, geometries, and layout. By allowing this, we see how much AH could be accepted by architects and professionals without imposing limitations that many times have been typical of others old approaches.

Buildings of the future, beyond 2020 and towards an equilibrium with nature are possible now, they are monitored and they are growing in numbers.

2.1.1 Experimental Houses at Politecnico di Milano: Prototypes and Monitoring Campaign

The UE has enhanced the environmental challenge since several years, accounting the 40% of overall energy consumption to the AEC sector and pushing to increase the energy efficiency of buildings and to lower our dependency on non-renewable sources. Buildings must become very energy efficient, with an annual balance between energy production and consumption close to zero: Nearly Zero Energy Buildings (2010/31/EU).

However, the current construction practice presents very often a big gap between the predicted and the real behavior of building, because of, of course, approximation

---

[1]Design and project by Università Roma Tre, team leader Prof. Chiara Tonelli, winner of Solar Decathlon 2014. Politecnico di Milano took care only of Active House simulations and labeling and then used that prototype as a case study for Mediterranean Active House in the Ph.D. studies by Arianna Brambilla, visible in next chapter of this book.

[2]Design and project by Atelier2—Arch. Valentina Gallotti (Architecture and Technology), VANONCINI SpA and Politecnico di Milano – Prof. Marco Imperadori (Engineering and development). Politecnico di Milano took care only of Active House simulations and labeling. The project is described in the chapter 4 of this book.

Fig. 2.1 VELUXlab, the first Italian NZEB located in a University campus (Brambilla 2013) (Ph. courtesy VELUX Group)

on simulation modeling approaches, but also users' interaction and bad use of these new smart objects.

In order to understand the magnitude of this gap and operate on its main causes, the ABC Department of Politecnico di Milano has been—and is currently—involved in several experiments on real building prototypes, 1:1 scaled: VELUXlab (Fig. 2.1) and C-ASA (Fig. 2.2).

At first, these two small buildings have been objects of an intense design phase, supported and validated by several modeling, analysis, and simulations. The design strategies were developed according to the AH principles, about indoor comfort, energy, and environment.

Afterwards, both the two buildings have been equipped with several different sensors and sensorized devices, in order to make real data available and to investigate the real operational phase of the two smart and active constructions.

According to this methodological approach, the process would become circular, based on input and feedback: the analysis and validation of the projects though different protocols and standards—the first one: AH Specifications—helps to understand the impact of initial design choices on buildings' behavior, and to develop further upgrades, with a special focus on Mediterranean climates. Then, the real-data collec-

Fig. 2.2 C-ASA is the Active House shelter of the Lecco Campus of Politecnico di Milano (Ph. Irene Giunta)

tion and mining process allows getting a concrete feedback through the monitoring campaign.

At the end, thanks to real affordable data, this approach allows to certify and finally reward the buildings with the AH label, in their "as-built" condition, closing the gap between expected and real performances.

2.2 VELUXlab

VELUXlab is the first prototypes of Politecnico di Milano and was established in the University campus of Bovisa—La Masa, in 2012. Since then, it is currently the object of several tests and scientific application, activities and experiments as a proper active lab (Fig. 2.3).

The laboratory was born in 2011, when VELUX and Politecnico di Milano started the renovation of the demo-house ATIKA, a model home previously designed by ACXT/IDOM as examples of innovative building for Mediterranean climates. The design team of Politecnico di Milano upgraded the construction: the implementation of the building envelope through multi-layered construction solutions improves its performances, while the plant system was completely designed to fit the purpose of the inside activities (Imperadori 2014).

This first phase of the design process was supported by static and dynamic analysis, aiming the minimization of energy needs (Imperadori et al. 2013; Brambilla et al. 2018).

Afterward, the building envelope and systems have been equipped, since the construction phase, with several sensors, interconnected via a wireless network, measuring:

- Temperature (°C): on the envelope as superficial and indoor air-temperature, with thermos-resistor sensors; on systems, as working temperatures;
- Humidity levels (%);
- Heat production;
- Electrical production;
- Electrical consumption;
- CO_2 concentration.

The huge amount of data, detected by those sensors and recorded in real time, are analyzed throughout a big-data mining process, in order to pattern the building behavior in its operational phase, with a focus on users' interaction.

During the experimental period, all the data related to the building characterization were translated into geometrical and performances information and uploaded on a BIM-based shared platform of augmented reality, following an IoT-driven approach (Fig. 2.4). This work targets the translation of the engineering knowledge to users-friendly information (Imperadori et al. 2018) (Fig. 2.5).

In order to attest the validity of AH principles also in the Mediterranean area, VELUXlab was elected as a case study of several research projects and a doctoral thesis (Brambilla 2014a, b). The high performances have been verified at first through

Fig. 2.3 VELUXlab is continuously the object of several investigations, according to the implementation of the knowledge on construction innovative materials and components; here a sample of the experiment that was conducted with a drone device, equipped with a thermos-camera in order to investigate the island urban effect and the thermal capacity of different materials (Brambilla and Imperadori 2013)

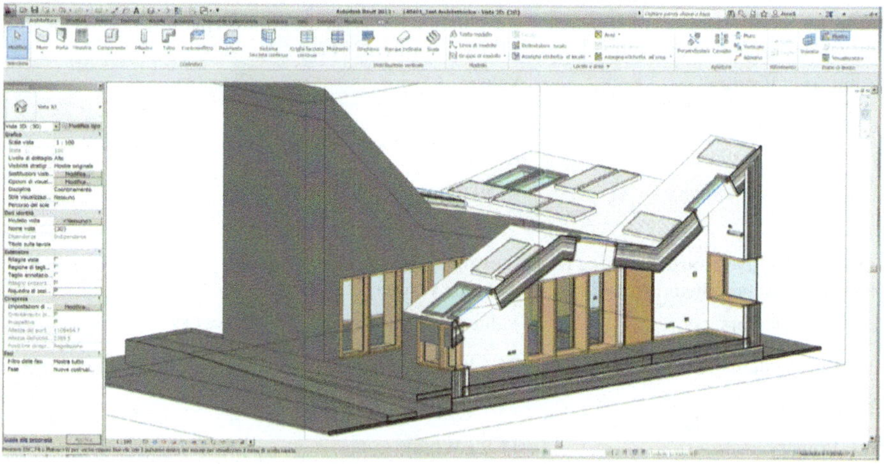

Fig. 2.4 The BIM project of VELUXlab is fulfilled with all the project information, reaching a LOD500, and available for querying on the shared-platform (Imperadori et al. 2018)

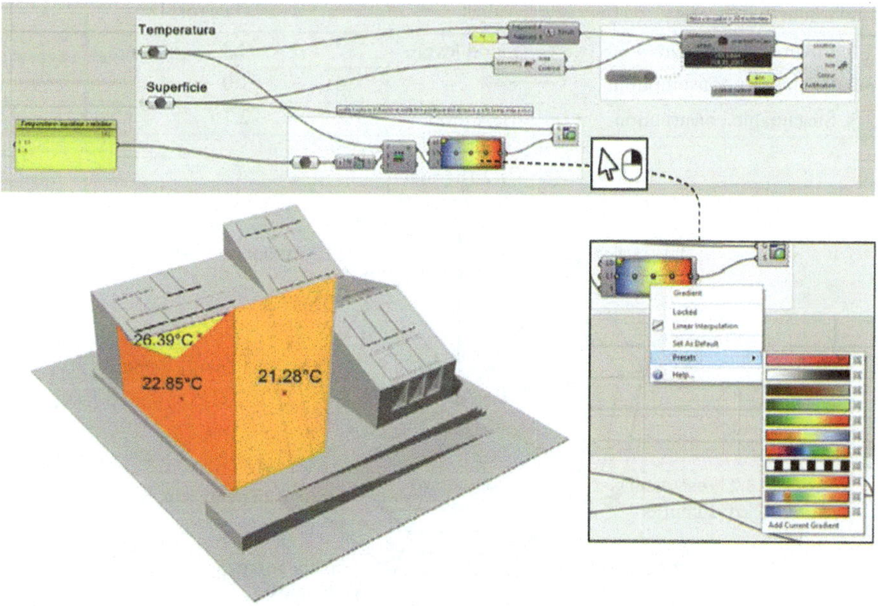

Fig. 2.5 Augmented reality and BIM model, implemented with big data from the sensors network (Imperadori et al. 2018)

the CENED protocol, where the building gained the A+ values, in 2014. The monitoring campaign headed also to achieve the AH label, as the first Italian Active House certified “as-built” (Tables 2.1, 2.2 and 2.3 and Fig. 2.6).

Table 2.1 Active House labelling results: evaluation levels for the comfort parameters (verification process performed by the Active House Alliance verifier, 2017)

Comfort indicators	Value	Category
1.1. Daylight	5.7%	1.0
1.2. Thermal environment	Better level	1.8
1.3. Indoor air quality	<750 ppm	2.0

Table 2.2 Active House labelling results: evaluation levels for the energy parameters (verification process performed by the Active House Alliance verifier, 2017)

Energy indicators	Value	Category
2.1. Energy demand	74.7 kWh/m^2	2.7
2.2. Energy supply	79.7 kWh/m^2	1.0
2.3. Primary energy	11.9 kWh/m^2	2.8

Table 2.3 Active House labelling results: evaluation levels for the environment parameters (verification process performed by the Active House Alliance verifier, 2017)

Environment indicators	Value	Category
3.1. Environmental loads	Good level	2.3
3.2. Freshwater consumption	84%	1.0
3.3. Sustainable construction	Best level	2.1

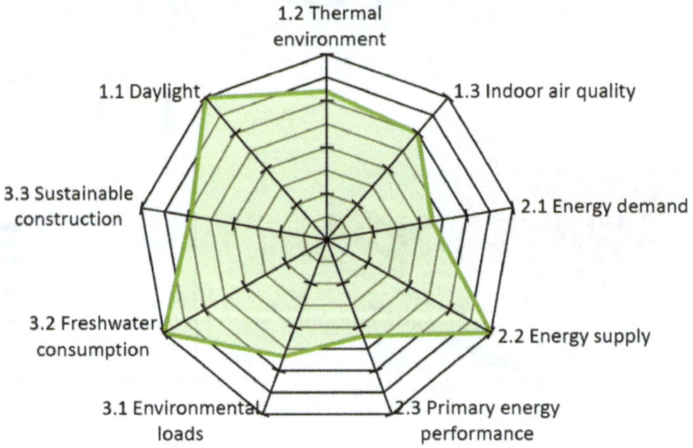

Fig. 2.6 Active House Radar of VELUXlab, evaluated by Politecnico di Milano design team and verified by the Active House Alliance, 2017 (© Politecnico di Milano and AH Alliance)

2.3 C-ASA

C-ASA (Construction-Advanced School of Architecture) is a small building prototype of Politecnico di Milano, located inside the campus of Lecco, hosting the new POLIMI Sport office (Fig. 2.7).

Its first architectural concept was developed within the first ASA (Advanced School of Architecture) Workshop, in 2015, where 27 selected students were asked to develop a concept for a new modular wooden-based unit partially supplied by an onsite energy production, through renewable source. Atelier2—Arch. Valentina Gallotti, Ri-Legno, Ing. Srl, and the design team of Politecnico di Milano have been responsible for the engineering process, selecting the technological solution for the envelope and defying the HVAC and PV systems (https://www.activehouse.info/).

This design phase has been supported by the verification of the building performances, in order to optimize the design strategies, lower the energy consumption and aim the indoor comfort. A set of dynamic simulation were performed, via TRNsys7 software, comparing different technological solutions and their performances in Lecco climate conditions (Fig. 2.8).

After the construction (Figs. 2.9 and 2.10), in order to validate the outcomes, several tests have been conducted, such as the blower door test (Fig. 2.10), and a monitoring campaign has been carried on for the first months of the building

Fig. 2.7 C-ASA is the building prototype of Politecnico di Milano which is located in the new campus of Lecco, to serve the research in the different field of the AEC sector and to host the POLIMI sport office (Ph. courtesy Politecnico di Milano)

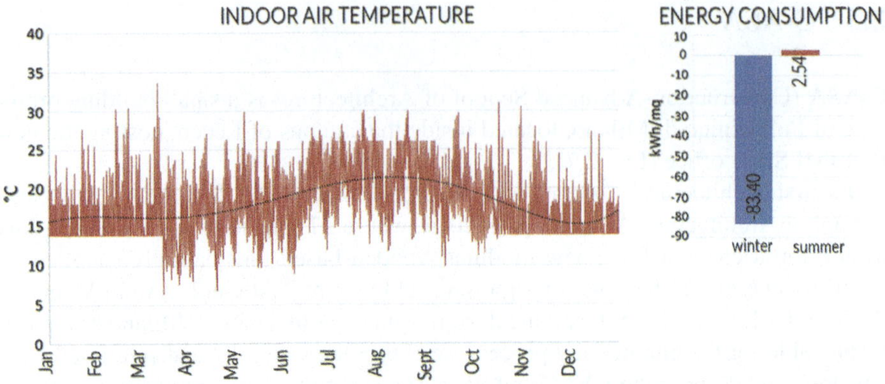

Fig. 2.8 Results of the dynamic analysis performed via TRNsys7 (Politecnico di Milano)

Fig. 2.9 The construction phase of the prototype (Ph. courtesy Politecnico di Milano)

operational phase. During the survey, the indoor space has been equipped with several temperature and humidity sensors, able to record a remarkable amount of data. The recorded values have been compared with the simulated ones, to validate the previous design phases. Finally, an acoustic campaign has been also performed, to test the performances of the envelope (Fig. 2.11).

Fig. 2.10 The construction phase of the prototype: the multi-layered hyper insulating sheet installation, by ACTIS (Politecnico di Milano)

After all this analysis, the building has been object of the AH evaluation, in order to obtain the AH label. The results are shown in the Tables 2.4, 2.5, and 2.6, and in the Fig. 2.12.

Fig. 2.11 Blower door test performed during the construction phase of C-ASA, in order to test the envelope performances (Ph. courtesy Politecnico di Milano)

Table 2.4 Active House labelling results: evaluation levels for the comfort parameters (verification process performed by the Active House Alliance verifier, 2017)

Comfort indicators	Value	Category
1.1. Daylight	5.8%	1.0
1.2. Thermal environment	Good level	3.0
1.3. Indoor air quality	<1000 ppm	3.0

Table 2.5 Active House labelling results: evaluation levels for the energy parameters (verification process performed by the Active House Alliance verifier, 2017)

Energy indicators	Value	Category
2.1. Energy demand	41.7 kWh/m^2	1.1
2.2. Energy supply	315.9 kWh/m^2	1.0
2.3. Primary energy	−596.5 kWh/m^2	1.0

Table 2.6 Active House labelling results: evaluation levels for the environment parameters (verification process performed by the Active House Alliance verifier, 2017)

Environment indicators	Value	Category
3.1. Environmental loads	Better level	2.3
3.2. Freshwater consumption	100%	1.0
3.3. Sustainable construction	Best level	2.1

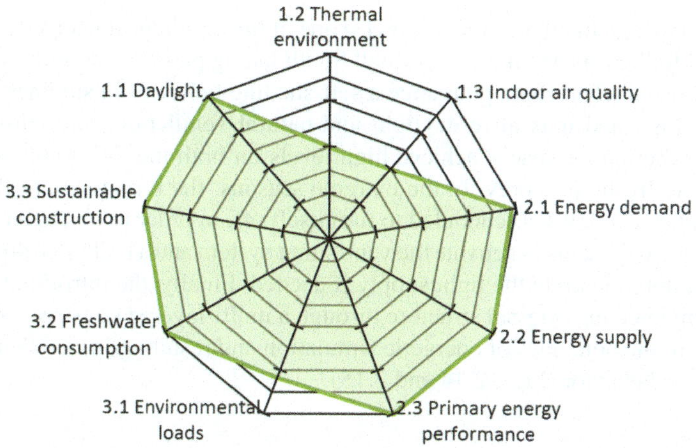

Fig. 2.12 The Active House Radar of C-ASA, evaluated by Politecnico di Milano design team and verified by the Active House Alliance, 2017 (© Politecnico di Milano and AH Alliance)

VELUXlab

Location	Milano, Italy
Project type	refurbishment
Use	office
Client	private company and public university
Design	Atelier2—Valentina Gallotti, Prof. Marco Imperadori – Politecnico di Milano
Year	2012

VELUXlab is the first Italian NZEB located in a University campus and the first Italian Active House "as built"-certified. It is an innovative laboratory of Politecnico di Milano, and it has been subject to several experiments about buildings' behavior, thanks to the evaluation of data of a monitoring campaign (Fig. 2.13).

Comfort
The retrofit design maintains the original shape of the building, conceived as a model home for Mediterranean climates: a small South facing patio, accessible from all the rooms, whose pitched roofs guarantee a self-shading behavior in summer and solar gains in winter and maximize daylight and natural ventilation, through skylights. Therefore, the indoor spaces achieve high levels on both the DF evaluation (up to 6%) and the fresh air supply. In the extreme seasons, the remote-controlled opening/closing of windows is scheduled to turn itself off, in order to obtain an adiabatic envelope; the CO_2 sensors activate the ventilating system and a LED self-dimmerable lighting system integrate the light supply if needed. Finally, the retrofitted envelope has implemented the original structure through a multi-layered construction system, according to the outcomes of energetic simulation and resulting in excellent thermal and acoustic behavior (Figs. 2.14 and 2.15).

Energy
The building design was conceived to answer instantly to climate changes and to naturally achieve indoor well-being, with the help of the mechanical system just in extreme seasons. It is equipped with an HVAC system, with heat recovery (90%), air-water heat pump (7 kWhp) and 3 solar panels with a 160l boiler for the domestic hot water supply. In order to partly cover the electric demand with renewable sources, a 2 kWp PV system has been install over the ventilated roof. The final energy demand is 11.9 kWh/m^{2}y. The efficient performances of the building's system are every year confirmed by the wireless sensors network monitoring system, that validates the building as a very smart and active center of research (Fig. 2.14).

Environment
The positive environmental footprint of the building depends on the use of innovative and recycled/recyclable materials, such as the recycled glass cladding panels of the ventilated facades. Moreover, the LCA has been performed for all the building elements, validating the building sustainability (Fig. 2.14).

Fig. 2.13 VELUXlab (Ph. courtesy VELUX Group)

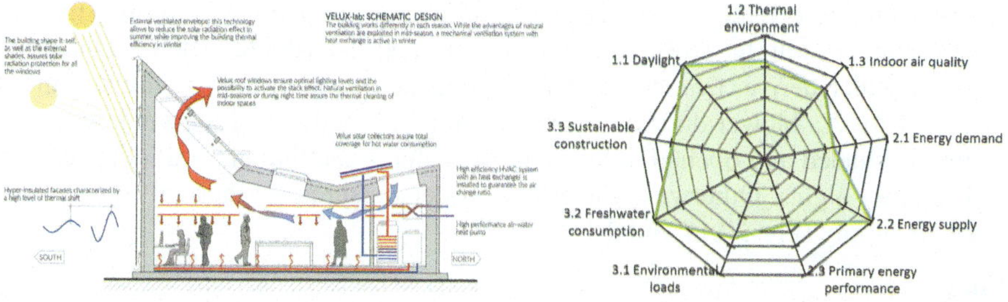

Fig. 2.14 VELUXlab schematic design and AH Radar (© Politecnico di Milano and AHAlliance)

Fig. 2.15 From the left to the right: as first, VELUXlab bright indoor spaces (Ph. courtesy VELUX Group); then, the adiabatic envelope (Ph. Marco Baccaro) and finally the pitched roof with skylights and solar panels (© Politecnico di Milano)

C-ASA

Location	Lecco, Italy
Project type	new construction
Use	office
Client	public university
Design Architectural design	ASA Students and ATelier2; **Engineering design** ATelier2, Ri-Legno, Ing Srl; **Developer** Actis SA, Alma Giorio, Cerlogne Pavimenti, Internorm Italia Srl, Knauf Sas, Rothoblaas Srl, Vanoncini Spa, Velux Italia, Zintek Srl
Year	2016

C-ASA (Construction—Advanced School of Architecture) is a small cubic shelter of Politecnico di Milano—Lecco Campus. It has a CLT structure, thermo-reflective aerospace-derived insulating layers, and outer Titanium zinc skin (Fig. 2.16).

Comfort
High performances materials and building components assure both the visual and thermal comfort of the 12.5 m^2 indoor space. In order to ensure the maximum inflow of natural light and avoid the glare effect, the envelope has a defined rhythm of transparent and opaque parts. Wide vertical windows, the glass door, and a skylight allow an excellent level of diffuse illumination and interesting views of the surrounding environment. Moreover, the 36 m^2 cantilevering roof protects the exposed facades from the direct sunlight. The layered envelope is made by a double-shell of insulating material, which derives from the aerospace applications, assuring a significant cut off of energy loss (a blower door test was performed). Finally, the roof windows could be totally shaded and opened, to activate the natural ventilation, in middle seasons and summer nights. Otherwise, the HVAC system ensures the thermal comfort and fresh air supply (Figs. 2.17 and 2.18).

Energy
The compact shape of the building, its orientation and the definition of fenestrations were modeled throughout energetic simulations, in order to optimize the energy balance all over the year. The massive CLT panel structure and the inner/outer insulating shells improve the thermal capacity to store energy during winter. Besides, the roof is added by an additional reflective insulation layer, in order to defend the building from the high solar radiation in summer. Finally, a solar canopy was installed nearby the building, to host 9 modules of PV panels (3 kWp), connected to the urban grid (Figs. 2.17 and 2.18).

Environment
The structure of the building consists entirely of CLT panels, which present the PEFC certification. Moreover, an LCA was performed to assure the reduction of the environmental load of the building. As a public office, it has no freshwater consumption (Fig. 2.18).

Fig. 2.16 C-ASA, the small shelter of Politecnico di Milano—Lecco Campus (Ph. Irene Giunta)

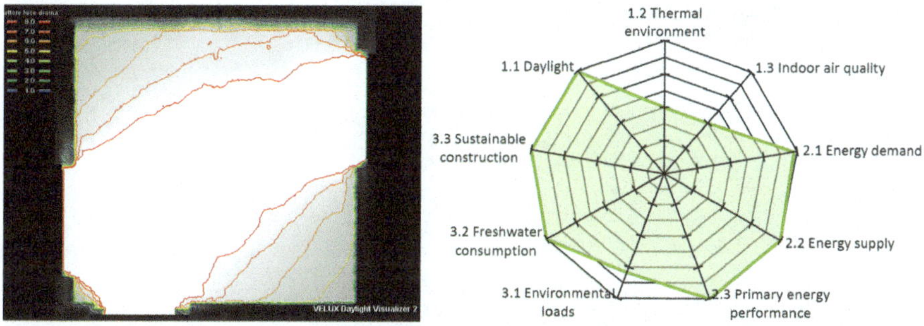

Fig. 2.17 C-ASA schematic design and AH Radar (© Politecnico di Milano and AH Alliance)

Fig. 2.18 C-ASA details: on the left, the CLT panels installation, under the Resegone majesty; in the center, the bright interiors and on the right, the small shelter at the entrance of the campus (Ph. Irene Giuta)

References

Brambilla A (2013) Comfort and energy assessment of the first Italian Nearly Zero Energy Building in a university campus. In: PLEA 2013—sustainable architecture for a renewable future proceedings, Munich

Brambilla A (2014a) First assessment on the principles for building a mediterranean Active House

Brambilla A (2014b) Mediterranean Active House: analysis of climate and users impact on energy and indoor comfort in the new efficiency standard for sustainable buildings, Politecnico di Milano, pp 71–76

Brambilla A, Imperadori M (2013) Velux Lab: innovation between art and science. In: Measuring the non-measurable future urban intensities proceedings, Tokyo

Brambilla A, Salvalai G, Imperadori M, Sesana M (2018) Nearly zero energy building renovation: from energy efficiency to environmental efficiency, a pilot case study. Energy Build

Imperadori M, Sauchelli M, Brambilla A, Falcone N, Zanello C (2013) Comfort and energy assessment of the first italian nearly zero energy building in a University Campus. In: Sustainable architecture for a renewable future

Imperadori M (2014) VELUXlab: a sustainable prototype of Active House for innovation. E3S Web of Conferences

Imperadori M, Salvalai G, Scoccia R, Brunone F, Fumagalli MA (2018) The sense of sensors. In: CIB 2018 Proceedings, Munich

The Directive 2010/31/EU of the European Par-liament and of the Council of 19 May 2010 on the energy performance of buildings, Official Journal of the European Union, 53, 2010

www.activehouse.info

www.Atelier2.it/opere/veluxlab/

www.Atelier2.it/opere/C-ASA/

Chapter 3
A Reflection on Active House in Warm Climates

Abstract Active House represents the next generation of sustainable buildings, which have a positive balance between benefits and resources consumed. In this chapter, the adaptation of the model for warm climates is explained. The aim is to investigate how the standard fits the Mediterranean region and how it can be used to design an Active House in hot climate. A sensitivity analysis on design parameters is performed to investigate their influence on energy consumption and thermal comfort provided according to the Specifications. RhOME, an outstanding example of efficient and sustainable design, is used to validate the analysis.

Keywords Active house · Sustainability · AH in hot climates
Energy efficiency · Thermal comfort · Design models · Model validation

3.1 Methodology for Southern European Climate

3.1.1 Active House Vision and the Mediterranean Region

Active House vision defines highly ambitious long-term goals for the future building stock: buildings that create healthier and more comfortable lives for their occupants without impacting negatively on the climate. The vision represents the next generation of sustainable buildings that encompass both energy and comfort. The Active House Alliance is a non-profit organization supported by a group of partners with the same goal: creating an independent and international alliance to promote this new balanced and holistic approach to sustainability throughout the construction sector. Active House approach relies on a clear and defined indexing of rules and suggestions, which are then transposed in the standard and visualized by the radar. The focus is not the idea of zero, but the intelligent use of resources coupled with an integrated approach to the whole building design. The radar is a decision-making tool: a compass that guides designers toward the optimization of indoor comfort, the maximization of energy efficiency and the reduction of environmental impacts.

This chapter is written by Arianna Brambilla.

L. Feifer et al., *Active House: Smart Nearly Zero Energy Buildings*,
PoliMI SpringerBriefs, https://doi.org/10.1007/978-3-319-90814-4_3

Table 3.1 Highlights of the main criticisms of the AHS parameters in relation to the warmer European region (2)

Area	Parameter	General criticism	Mediterranean criticism
Comfort	Daylight factor	the score is set by the room with the lowest daylight factor. It does not allow exceptions	having high DLF means having big openings to light and direct solar radiation
	Direct sun availability	the score is set by the room with the lowest factor. It does not allow exceptions	having direct solar radiation inside the rooms means also increased internal solar gains and risk of overheating
	Minimum operative temperature	Not clear range of validity of the temperatures intervals The score is set by the room with the lowest value	The evaluation is made on the system's possibility and not reality
	Maximum operative temperature	Not clear range of validity of the temperatures intervals The score is set by the room with the lowest value	The evaluation is made on the system's possibility and not reality
	Standard fresh-air supply	The score is set by the room with the lowest value. The evaluation does not take in consideration different size/destination of the spaces	It implies a strong use of natural ventilation that brings into spaces the hot air from outside
Energy	Annual energy demand	Different ranges should be set for different buildings typologies: renovation versus new, residential vs office, apartment versus single family dwelling	In the calculation cooling calculation is simplified compared to heating
Environment	Minimization of fresh-water use	not clear definition of average use of water per building per year. Parameters evaluated through comparison with a "standard" building use	Freshwater saving is already deepen in Med area, the improvement is reductive and misleading

The Active House Specifications were developed through an open-source model, involving online debates and contributions as well as offline meetings and workshops with broad participation across the Alliance partners. The process relies on the empiric results of several case-studies; thus the outcome is strongly connected to buildings analysed. The Alliance was born in the North Europe and all the research partners were from this region, until Politecnico di Milano joined it. Politecnico di Milano took the responsibility of validate the Active House Specifications for the southern Europe, characterized by a warm to hot climate.

Table 3.1 summarizes the criticisms identified during the scoping phase of the Mediterranean-Active House research project. Among all, direct sun availability, operative temperatures and energy demand have been considered the most influential for the success of the vision in hot climates.

3.1.2 Mediterranean Region?

The Mediterranean-Active House research is focused on the adaptation of the vision to the warmer South Europe, but what do we mean exactly with "warmer"?

Mediterranean climate presents different and complex climatic conditions. It is weather stratified, extremely variable and it has not a predominant thermal flux direction (e.g. heating for northern region). Winter is not rigid but it requires artificial heating while summer has high cooling loads, resulting in high energy consumption for conditioning the indoor space.

It is important to locate this region: it is clear that it is an area on the South of Europe, located around the Mediterranean Sea, but where are its boundaries?

In order to have a more precise structure of the European climatic zones -and so a precise idea of what it's meant with cold, hot and warm climate- it is necessary to create a grid throughout the Continent.

The methodology used to create the climatic matrix consists in overlapping different weather maps characterized by different clustering parameters or precision levels. The zoning process is usually made on three important factors: average temperature in the coldest month, average temperature in the hottest month and number of days with a temperature below 5 °C or above 25 °C. However, it is possible to make another classification based on the parameters that affect most the energy consumption of a building (Erell et al. 2003). These are:

- temperature: probably the most known parameter but also the most effective in building design. In order to provide a meaningful representation of the outdoor temperatures progression, at least two values are needed: the daily average and daily amplitude, or the maximum and minimum values. In this study, the first couple is used for the zoning and the second for finding the reference cities (a city that can be used as representation of the climatic zone);
- solar radiation: generally measured on horizontal surface in its two components: direct (beam) and diffuse. Beam radiation is responsible for heating the spaces, it must be shaded in summer to prevent overheating and enhanced in winter for passive heating. Diffuse radiation is related to the illuminance and can increase critically the indoor glare. For weather clustering, solar radiation is often associated to cloud cover fraction, precipitation or number of rainy days;
- humidity: although it does not affect directly the buildings energy consumption, it is important in human thermal comfort balance. In this analysis, the maximum and minimum relative humidity are used to identify the reference cities.

In the Mediterranean-Active House research, the first step of the climate clustering analysis was represented by the Koppen climate classification review (Rubel and Kottek 2010); Koppen made complex system of classification based on empirical measures, where for each climate is given a climatic formula, defined upon statistical operation related to average monthly temperature, rainy days and annual precipitation. Koppen classification divides Europe in five main categories: Nordic from Iceland to Urali, Oceanic, Alpin for the highland, Mediterranean and Continental.

Although it is widely used by botanist and bio-scientists, it is based on a climatic function of only precipitation and average temperature, which makes it not suitable for the building design use. In order to refine the classification, we integrated the formula overlapping the climatic zoning based on other three parameters: monthly

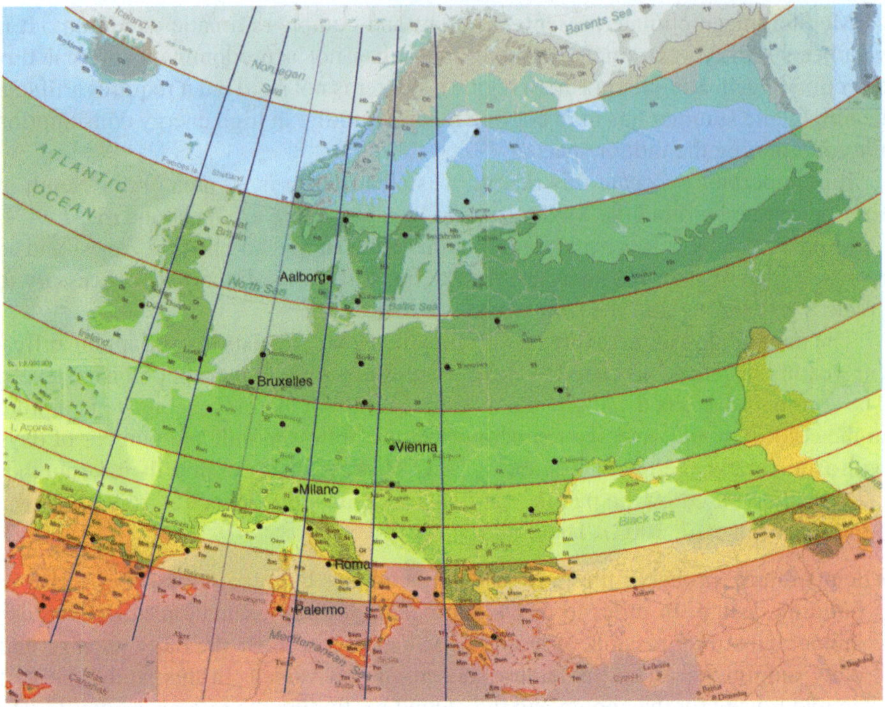

Fig. 3.1 Definition of different climates clusters in Europe and reference cities that represent the climatic conditions of the clusters

amplitude of temperature, monthly average and maximum/minimum value of solar radiation (total). Every comparison allowed to tighten the climate areas and better define their boundaries.

According to the method described, it is possible to divide Europe in six different climates, the most relevant for the study proposed are the two Mediterranean and the pre-Mediterranean one. The difference respect Koppen classification is basically the sub-division of the Mediterranean climate into two different categories, the addition of the pre-Med area and the detailed division of the Continental one. These clusters are weighted on the radiation, which influences the climatic conditions (Fig. 3.1).

Building simulations request a specific weather file, representing a specific location. For this reason, it is necessary to define a reference city that could well summarise the characteristics of its weather cluster. Thus, after the definition of the climatic zones, we performed an analysis aimed at the identification of the reference city for each of these zones. For each weather cluster, different cities have been analysed in relation to their climatic parameters in both summer and winter, using a hierarchical clustering of variables, assuming that the inclusion of multiple and collinear factors in the analysis affects its results (Mimmack et al. 2001). Four significant physical quantities have been used for each season:

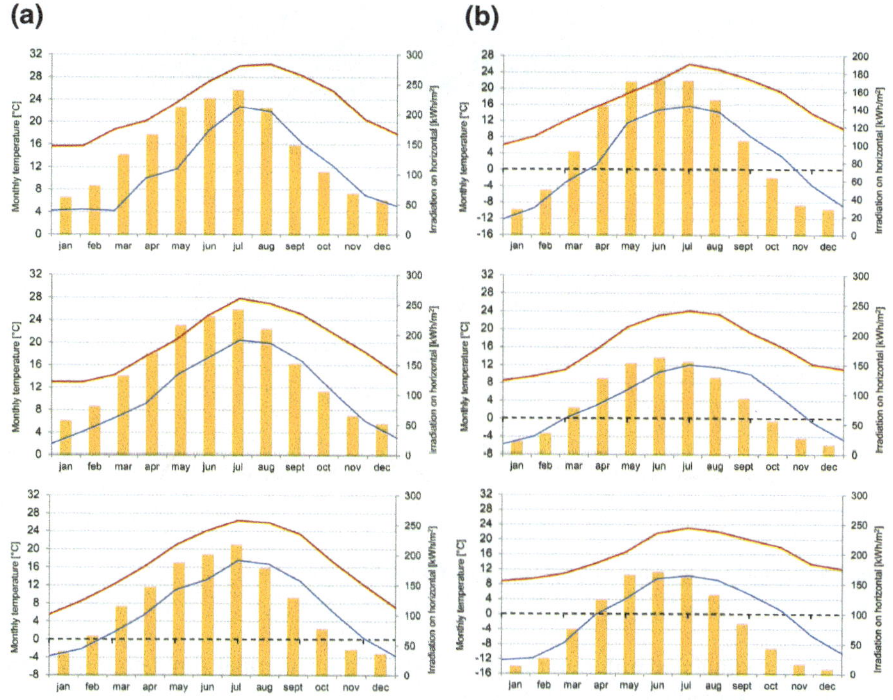

Fig. 3.2 Climatic condition of the reference cities

summer: mean daily temperature, daily amplitude of temperature, solar radiation and daily average humidity;
winter: mean daily temperature, daily amplitude of temperature, solar radiation and precipitation factor.

The reference cities chosen to represent the climatic clusters are: Palermo (cluster1: very hot), Rome (cluster 2: hot to warm), Milano (cluster 3: hot to cool), Wien (cluster 4: warm to cool), Uccle (cluster 5: cool to cold), Aalborg (cluster 6: cold). The main average monthly temperature and horizontal irradiation are shown in Fig. 3.2.

3.1.3 A Model of Mediterranean Active House

In order to assess the reliability of Active House as sustainable standard in the Mediterranean region, the first step is the analysis of very efficient buildings already built in the area and creating a basic model, sum of all the common features.

We firstly performed a preliminary analysis to understand the influence of the basic space configurations on the annual energy performance of a simple house design.

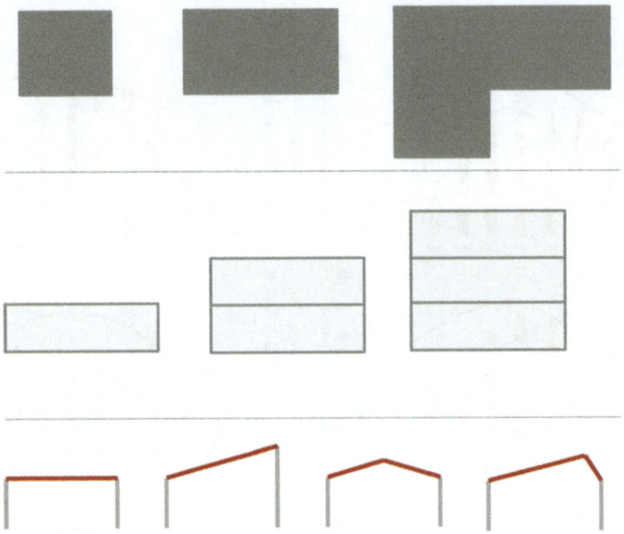

Fig. 3.3 Example of the options catalogue used for the first analysis

To do so, we simulated different models in the dynamic thermal simulation software Trnsys. The assumptions used (Fig. 3.3):

- FIXED for all cases—Envelope as a mixed-technology: massive floors and light walls
- FIXED for all cases—Single residential unit
- CHANGING parameters—floor area, number of floors, shape of the roof and articulation of the space.

The most interesting results are related to the exchange surface: the bigger is the wall area, the more heating is required. At the contrary, big roof surfaces exposed to south mean higher cooling needs (almost the double if compared to not south-exposed roof). These results highlight the importance of the shading strategy and the necessity to minimize the surface (both transparent and opaque) directly exposed to solar radiation.

From this very first round, it has been possible to create a building geometry that included the most advantageous shape's combination. Based on this, we integrated the model through architectural articulation, technological refining and technical implementation based on a database of very efficient buildings located in the climatic zones identified as cluster 1, cluster 2 and cluster 3. The selection of the case-studies encompassed shape decision, materials selection, energy demand management and

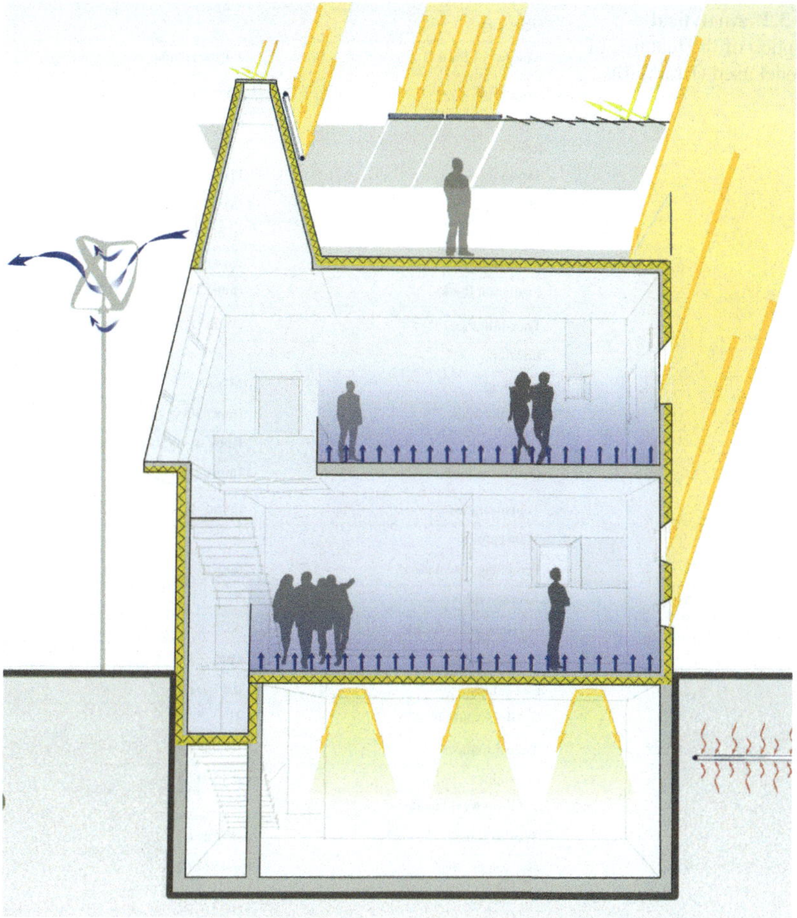

Fig. 3.4 Model of Active House for warm climates: opened to north to capture cold light and closed to south to protect from the direct solar radiation; integration of ventilation stack. Use of the renewable sources available to produce energy, such as sun (PV and solar), wind (windmill) and ground (geothermal pipes)

optimization of the solar technology performance could enhance the social, economic and environmental sustainability of a building (Miller and Buys 2012); and it included built examples (Brambilla 2014), as well as empirical research results (Nicol and Pagliano 2007; Baker and Standeven 1996; Voss et al. 2012; Athieniti and Pagliano 2010; de Graca et al. 2012; Sartori et al. 2010).

The results of this first design step is the Basic Model of Active House for warm climates (Fig. 3.4; Table 3.2).

Table 3.2 Analytical description of the features of the model used (Brambilla 2014)

Geometry	
Number of floor	1 + mezzanine and ground
Area	(m^2) 97
S/v ratio	0.8
Roof	
Typology	Plane shaded
Mass effect	(h) 10
Total thickness	(mm) 280
Insulation	wood wool
Insulation thick	(mm) 250 (+100)
Transmittance	(W/m^2K) 0.117 (0.085)
Envelope	
Typology	Mixed
Total thickness	(mm) 300
Insulation	Rock wool/polyurethane
Insulation thick	(mm) 240 (+100)
Transmittance	(W/m^2K) 0.137 (0.077)
Transparency	
Percentage on surface	40
Average fld	5
% Glass-north facade	35
Transmittance	(W/m^2K) 1.1
G–t values	0.45–0.67
% Glass-south facade	10
Transmittance	(W/m^2K) 1.1
G–t values	0.45–0.67
% Glass-west facade	5
Transmittance	(W/m^2K) 1.1
G–t values	0.45–0.67
Energy system	
Ventilation	Hybrid
Ventilation ratio	4
Hr efficiency	(kJ/hK) 1000 (1500)
Renewable sources	Solar–PV–geothermal
Production	Heat pump
Other	Stack ventilation

3.1.4 Sensitivity Analysis: A Tool to Understand the Most Influential Parameters for Sustainable Design in Warm Regions

Sensitivity Analysis helps to understand the correlation between the uncertainty in the output of a system and the uncertainty in its input. It is a method to study both qualitatively and quantitatively how much a variation in the model assumptions can influence the output.

In the design process, it can be helpful to inform the designers about the parameters that mostly influence the final energy performance or the indoor comfort provided, optimizing the process and focusing on what really drives those performances.

In relation to the design of innovative buildings a sensitivity analysis can be useful for:

- robustness of a design criteria to different scenarios enhancement;
- correlations clarification between input and output in a fixed system;
- uncertainty reduction and robustness growth;
- model simplification through the elimination of ineffective parameters;
- model optimization.

Monte Carlo Analysis is a statistical method that relies on repeated random sampling to obtain the distribution of an unknown probabilistic entity. Usually it is articulated in five consecutive steps:

1. selection of design criteria and design parameters with their distribution;
2. creation of a sample matrix of the parameters;
3. evaluation of the model for each set of the matrix;
4. uncertainty analysis;
5. sensitivity analysis.

There are several methods that can be used to perform a sensitivity analysis (Chan et al. 1997); the most common is the one-factor-at-time changing approach (OAT). Among all the OAT approaches, Morris Method is the most suitable for our purpose, as it can handle hundreds of input factors without relying on strict assumption (Campolongo et al. 2010). The potential of this method is bigger the earlier is applied in the design process (Heiselberg et al. 2009).

In the Mediterranean-Active House research, we performed the sensitivity analysis considering three different design criteria: minimization of cooling needs, minimization of heating needs and minimization of annual energy consumption.

The design parameters considered (Brambilla 2014):

- Wall Transmittance (value steps: 0.06–0.085–0.137–0.2–0.3–0.45 W/m^{2}K, symbol: U);
- Windows Transmittance (value steps: 0.7–1.1–1.4–2.8–4.6 W/m^{2}K, symbol: Uw);
- Heat Recovey Efficiency (value steps: 0–500–1000–1500–2000–2500 kJ/hK, symbol: N);
- Ventilation Ratio (value steps: 1–2–3–4–5 vol/h, symbol: V);
- Transparency Distribution Difference (value steps: 0–0.2–0.4–0.6–0.8–1, symbol: D):
- Mass Effect (value steps: 4–6–8–10–12–14 h, symbol: M).

The factorial sampling method proposed by Morris was used to generate the input matrix (Morris 1991). The method generates an OAT matrix, which contain the different configurations of the parameters chosen that must be assessed. We simulated the different configuration using the software trnsys, which allowed us to calculate the cooling, heating and annual energy consumption for each case. The results of the simulations have been assessed with the Elementary Effects Method (Saltelli et al. 2000).

This method allows to calculate the mean value and the standard deviation of each parameter on the final criteria used in the analysis, in our case: heating, cooling and annual energy consumption. Calculated these values, it is possible to visualize the results on a simple graph that indicates the relation between changes in each parameter and their effect on the criteria. Analysing the position of the points representative of the design parameter in the chart:

- if it falls at the left of the line: the correlation with the design criterion is non-linear
- if it falls on the right: the correlation is linear
- it is very closed to the line: the correlation is a mix of the two.

Moreover, the distance from the line describes the intensity of the correlation. The following graphs show the sensitivity analysis results for the three Mediterranean climates; assessed for heating reduction, cooling reduction, total energy reduction (Fig. 3.5a–c).

The results obtained clearly show which parameter has more influence on the energy performance and they represent a powerful tool to focus the design choice on the effective strategy to increase the energy efficiency. The charts inform about the possible consequences of changing a design parameter.

Table 3.3 show the consequences of changing the value of the design parameters. The table indicates it causes a positive or negative effect, and its intensity.

An example of how to read this table is represented by the walls transmittance (U value). In climate 2 increasing the U value reduces the heating consumption,

Table 3.3 Description of the influence of each design parameter on the design criteria (cooling needs, heating needs, total energy needs)

Climate	Parameter	Heating	Cooling	Total
1	U	Not influent ×	Increase ++	Increase +
1	Uw	Decrease –	Decrease –	Decrease –
1	N	Not influent ×	Not influent ×	Decrease –
1	V	Not influent ×	Not influent ×	Not influent ×
1	M	Not influent ×	Decrease –	Decrease –
1	D	Increase +	Decrease –	Decrease –
2	U	Decrease –	Increase +	Increase +
2	Uw	Decrease –	Decrease –	Decrease –
2	N	Increase +	Decrease –	Not influent ×
2	V	Increase +	Increase +	Increase +
2	M	Not influent ×	Decrease –	Decrease –
2	D	Increase +	Decrease –	Not influent ×
3	U	Decrease –	Increase +	Decrease –
3	Uw	Decrease –	Decrease –	Decrease –
3	N	Decrease –	Decrease –	Decrease –
3	V	Increase +	Not influent ×	Increase +
3	M	Not influent ×	Decrease –	Decrease ×
3	D	Increase +	Decrease –	Increase +

Effects are evaluated for a unit increment of the parameter considered

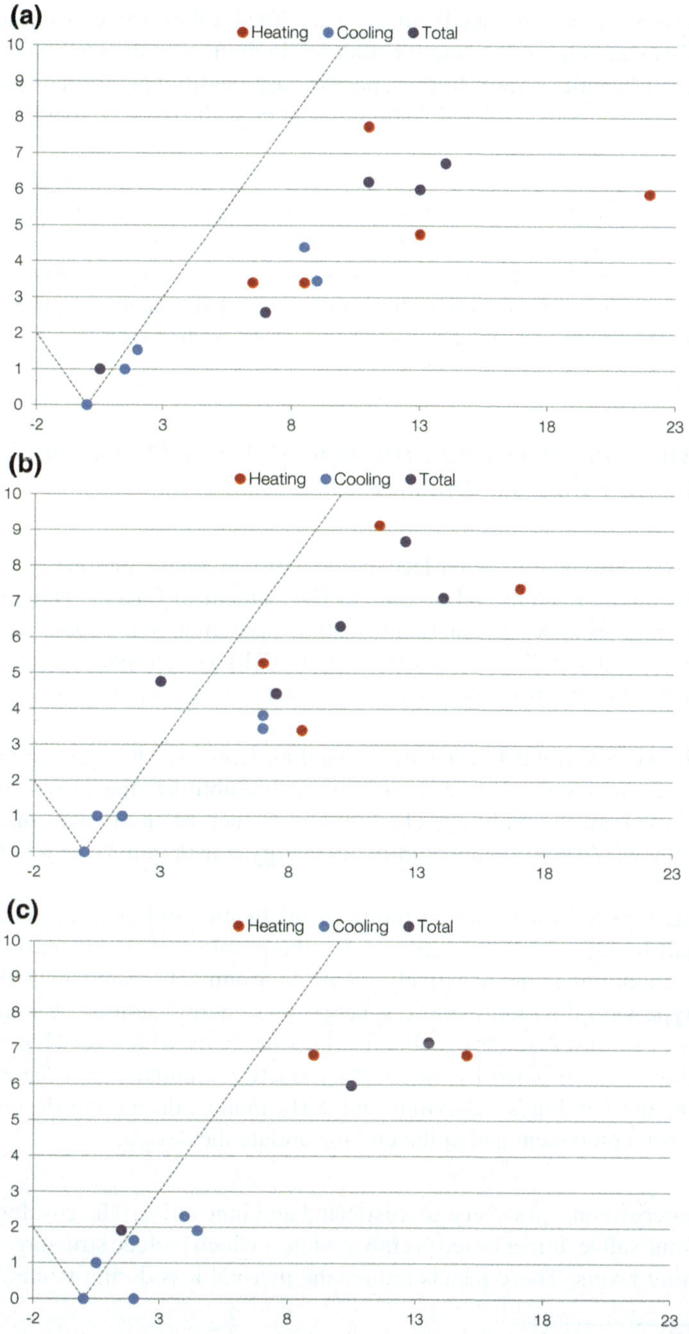

Fig. 3.5 **a** Resulting graph of the sensitivity analysis—climate 1. Uw is out of the area (high linear correlation). **b** Resulting graph of the sensitivity analysis—climate 2. Parameters not displayed are overlapped on the Cooling V factor (0, 0). **c** Resulting graph of the sensitivity analysis—climate 3. Total and Heating U and Uw are out of the graph area, showing high linear correlation

but it increases the cooling needs. In this case, the total energy consumption is the indicator that can help in deciding whether the increment in the walls transmittance brings overall benefits or not. In this specific case, additional thermal resistance in the envelope is not beneficial and it should be kept as the basic Active House model proposed.

In climate 1 and climate 2, where the energy demand is led by cooling needs, the most important factor to evaluate is the mass effect of the envelope. This highlights that to guarantee energy efficiency in these climates it is essential to calibrate the massive behaviour of the envelope and to improve the ventilation system. In climate 3, instead, it is possible to notice the importance of the heating season: the total demand is influenced mostly by the whole envelope performance.

3.1.5 An Award-Winning Prototype of Active House for Warm Climates: RhOME

RhOME[1] was designed for Solar Decathlon 2014: an annual international competition for Universities, promoted by the US Department of Energy. The competition asks to the students to design and build an innovative small living space that could be an example of sustainability and smart living. In 2014 the European competition took place in Versailles, France, with only twenty selected University teams competing (Fig. 3.6).

RhOME won the competition with outstanding score in all the categories. RhOME is designed following the Active House Protocol to optimize the design solutions in all the life cycle of the building. The building is seen as an efficient machine that optimizes the whole comfort and minimizes energy consumption (www.activehouse.info).

RhOME aims to be a model for future intensification of big cities: it is an example of small living module that could solve the problem of the increasing housing demand without impacting negatively on the environment. Team RhOME designed the prototype with the motto "more synergy, less energy" aiming at a harmonious system envelope/installations capable to works perfectly with a very low amount of energy. The design was led by different parametric simulations: at the early stage to optimize the building's behaviour and performance, during the design stage to calibrate each component and at the end to validate the design.

Energy

RhOME energy concept is very sophisticated and innovative. The envelope is composed by innovative timber based prefab modules, which include structure, insulation and finishing layers. These panels reduce the thermal loss during winter, thanks to

[1] Architecture, engineering, and development of University of Roma Tre—Prof. Chiara Tonelli and Prof. Stefano Converso—for Solar Decathlon 2014; Active House validation Politecnico di Milano—Prof. Marco Imperadori, Prof. Gabriele Masera, Eng. Graziano Salvalai and Eng. Arianna Brambilla. The project is resumed also in the following data sheet.

Fig. 3.6 RhOME, main façade and entrance, picture taken during the competition (*credits* L. Procaccini, team RhOME)

the high insulation levels, and smooth the heating peaks during summer, thanks to the high thermal mass and the ventilation cavity behind the external cladding.

RhOME has big openings towards south, to enhance passive heating during winter, while the big loggias shade them during summer, preventing the unwanted solar radiation and reducing the overheating risk.

RhOME is designed with a great attention to passive design and it relies on bioclimatic strategies to minimize the energy needs (Fig. 3.7).

After the attention on reducing energy needs, the following step is producing energy in a smart way. RhOME has a very intelligent conditioning system, that uses a heat pump in combination with thermodynamic panels, integrating the DHW needs and the ventilation system. This system is very innovative because it can produce hot water using only air temperature, every day of the year, thanks to the negative temperature of the fluid in the pipes. PV panels are innovative too: the system used is a Flexible Photovoltaic Panel, tiltable and slidable, used also as a curtain wall to shade the loggia while producing energy (Brambilla et al. 2017).

The total energy consumption in Contest Week is very low: only 3.48 kWh/mq. The different percentage for use are:

- Cooling = 55.50%
- Ventilation = 0.19%
- Domestic Hot Water = 7.43%
- Lighting = 1.85%

Fig. 3.7 Innovation in the energy supply system: PV system as curtain wall and solar thermal panels as baluster (*credits* L. Procaccini, team RhOME)

- Appliances = 26.60%
- Monitoring = 8.41%.

Environment
Team RhOME focused on the minimization of the building's environmental impacts, reducing the embodied energy and the life-cost of the materials. RhOME is built with only recycled or natural materials, which imply a reduction of the grey energy for each material up to 30% compared with the product as new. The approach used is called cradle-to-cradle, as basis of a circular design method, which minimizes the waste and assure a second life to buildings materials and components.

The most innovative strategy to reduce the environmental impacts is represented by the massive layer: it is made of recycled sand taken from the waste the building's excavation site. In a broader context, team RhOME thinks that the same approach can be used in cities, taking the sand from planned demolition interventions; demonstrating that everything can be a resource in a smart design for sustainable buildings.

Comfort
RhOME is designed to maximize the indoor comfort and to create a good experience for users. Big openings towards north and south enhance natural cross ventilation to reduce summer internal temperatures and assure good fresh air changes. Moreover, the massive layer shifts the thermal peaks smoothing the temperatures in summer, and stores internal heat in winter during the day and releases it during the night, functioning as thermal balance (Fig. 3.8).

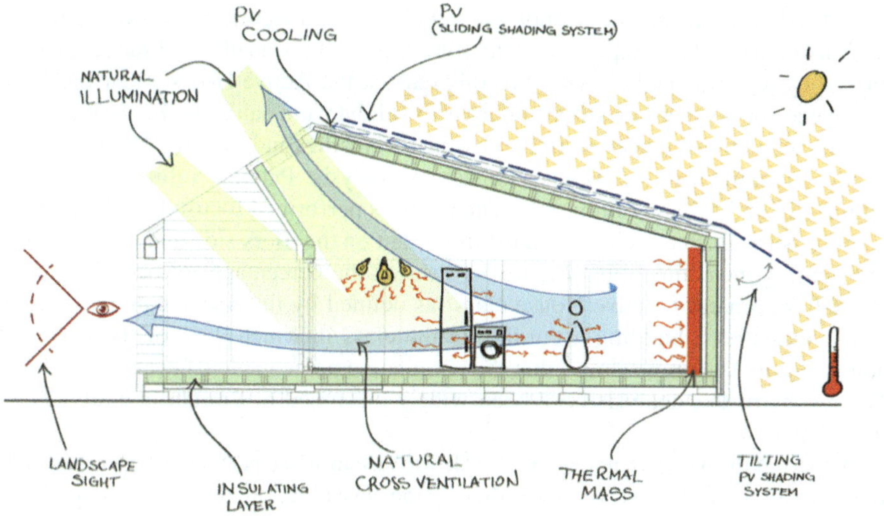

Fig. 3.8 Indoor environment inside the loggias (*credits* L. Procaccini, team RhOME)

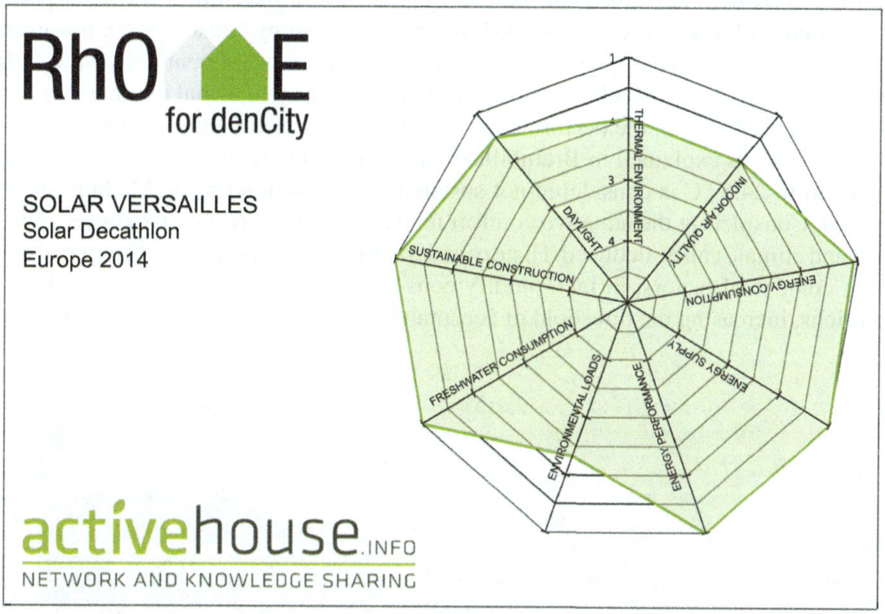

Fig. 3.9 Active House radar for RhOME (*credits* L. Procaccini, team RhOME)

Daylight is maximized by the roof windows facing north, catching the cold natural diffuse light that creates a comfortable and liveable indoor environment for users (Fig. 3.9).

RhOME was designed to respond to hot climates, however, it needed to perform well also during the competition, which took place in Versailles—France, characterized by a continental climate. For this reason, the Active House radar has been calculated for different locations, in order to validate and calibrate the design.

The calculation has been made in relation to Versailles and the three Mediterranean climates identified in the weather clustering analysis: Palermo (climate 1), Roma (climate 2) and Milano (climate 3). The analysis performed focused on the thermal comfort performance, as it is the most important on the users-side and the hardest to fulfil during the competition. Figure 3.10 shows the percentages of occupied hours that fall within each Active House class (as defined by the Specifications). Active House classifies a building according to the class in which 95% of the occupied hours can fit. The graph shows that the cumulative percentage is never above three. The results of this analysis have been widely discussed in Brambilla et al. (2017) (Fig. 3.10).

As part of the Mediterranean-Active House research, we performed a second analysis in order to understand what was driving the classification. Set point temperatures were modified progressively, testing the effects on the radar. During the procedure, it has been possible to observe that the progressive raise of the comfort class was not mirrored by a decrement of the energy indicator: a difference in the set point temperature of 1 degree corresponded to double the energy consumption, however, the building is very energy efficient and the difference is as influential as we would have expected. From this observation, it is possible to conclude that thermal comfort is the driving factor in the performance assessment for energy efficient buildings.

However, as explained in Brambilla et al. (2017), "the reduction of the set point from 26 to 24.5 °C is considered not sustainable as solution for the Mediterranean region. Considering the adaptive comfort model, a strong correlation between comfort and climate can be deduced. History and tradition of a population, in conjunctions with contextual factors, in fact, modify consistently thermal expectations and perceptions, increasing the threshold of acceptability of indoor temperatures in warmer

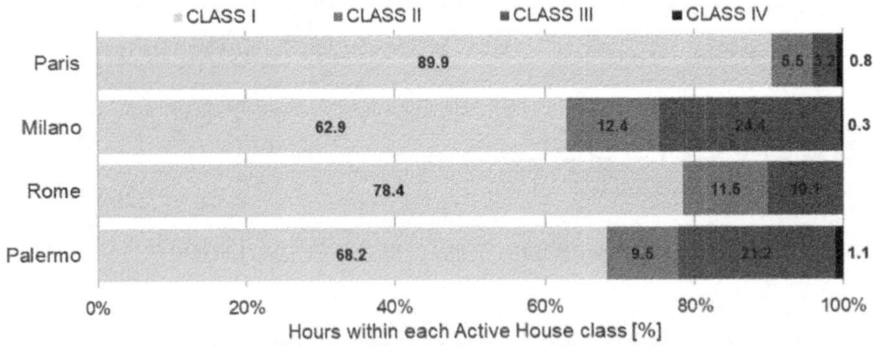

Fig. 3.10 Percentage of hour in each class of comfort. The A score is achieved with the 95% of hour in the class (red line). Winter comfort above and summer comfort below (Brambilla et al. 2017)

areas and decreasing it in colder regions (Van Hoof and Hensen 2007; IPCC 2014). The AH thresholds for the first two comfort classes are really restrictive and almost unusual for warm climate, since that these are translated into operative indoor temperatures lower than 25 °C (or 26 °C for the second class). A new ranking threshold has been introduced to acknowledge for the acclimatization opportunity, the Italian normative and the results highlighted in the previous analysis (Table 3.4)."

A second assessment was then performed according to the new thresholds. Figure 3.11 shows that the continental climate is the only one to score below the second class. The cause is probably the free cooling strategy adopted by RhOME during the middle seasons. However, the results show that a slight variation into the classification system is translated into big differences on the results, and this new system seems to fit better the Mediterranean building and living culture (Fig. 3.11).

Table 3.4 Proposed thresholds for the evolution of thermal comfort in warm climates, where A* is equal to 0.33 Trm

Proposed comfort set-point	
Maximum operative temperature (T_{op}) applied when $T_{rm} > 12$ °C, evaluated through hourly T_{op}, requirements should be met for at least 95%	With cooling system 1: $T_{op} < 26.0$ °C 2: $T_{op} < 26.5$ °C 3: $T_{op} < 27.0$ °C 4: $T_{op} < 28.0$ °C Without cooling system 1: $T_{op} < A^* + 21.8$ °C 2: $T_{op} < A^* + 22.0$ °C 3: $T_{op} < A^* + 22.8$ °C 4: $T_{op} < A^* + 22.8$ °C

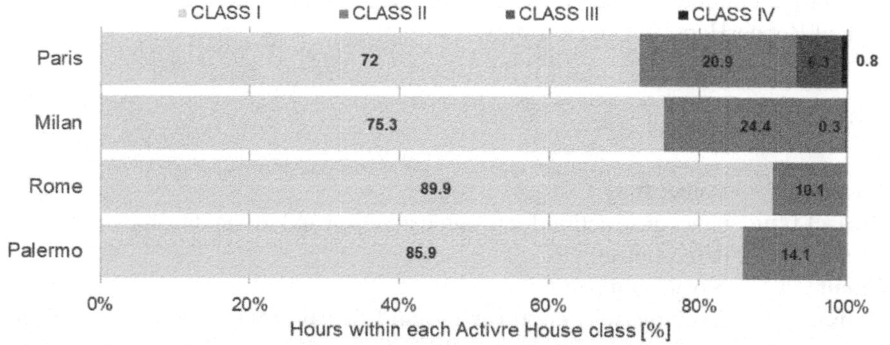

Fig. 3.11 Percentage of hour in each class of comfort. The A score is achieved with the 95% of hour in the class (red line). Winter comfort above and summer comfort below (*credits* Brambilla et al., 2017)

3.1.6 Lessons Learned

RhOME won the Solar Decathlon Competition 2014 thanks to its innovative approach to design: the radar, used as a decision-making tool, helped to reach an important result in an International context. Sustainability is not just a label but it is a holistic approach capable to empower the building's performances.

RhOME is an example of very efficient design: resilient to small changes in climatic conditions but reliant on strong bioclimatic assumptions: comfort results in weather clusters 1, 2 and 3 are optimal, but the same design is not valid for cool climates, such as Paris. Mediterranean- Active Houses are designed for a different climatic and cultural context and they should also be evaluated accordingly. The proposed new threshold modifies only the upper limits, yet the results are completely different. The standard should be able to acknowledge the social, cultural, economic and climatic difference throughout Europe.

The methodology used to define the basic model of Mediterranean Active House can be a perfect example of informed design; while the sensitivity analysis is a powerful tool to guide designer in focussing on the parameters that really influence the building performances in warm climates.

The case study shows that design criteria influence each other, often in a contradictory way: improving comfort can lead to higher energy consumption. Thus, the simultaneous validation of the different performances is essential to guarantee efficiency through the whole building life.

Multi-dimensional approaches, such as Active House, are the next generation of sustainability standards as they are able to guide designers toward the whole building efficiency, meant as reduction of the energy used, reduction of the resources involved and maximization of users' experience.

Now the building prototypes has been located as a demo-house in the exhibition park of RubnerHaus, in Chienes, Italy.

RhOME

Location	Rome, Italy
Project type	new construction for volume addition and urban densification
Use	demo-house
Client	private company
Design	**Architecture, engineering, and development** University of Roma Tre for Solar Decathlon 2014; **Active House validation** Politecnico di Milano
Year	2014.

RhOME is the winning prototype of the Solar Decathlon 2014. It represents an affordable example of Active House application to the Mediterranean climate. Moreover, it has been proposed as an innovative solution for the urban densification, being installed over the existing buildings' roofs: RhOME for DenCITY (Fig. 3.12).

Comfort
RhOME is designed for its occupants' well-being. The big openings to south and north enhance the cross natural ventilation, that positively prevents overheating and assures thermal comfort in middle seasons. A massive layer is used to mitigate temperatures in extreme seasons, thought the thermal shifting effect in summer and the heat storage in winter. Then, the north-facing roof windows and the exposure of the second loggia supply cold natural light, to achieve an optimal visual comfort. About the air-quality, the systems actively reduce PM10, CO_2 and VOC levels, inside the building (Figs. 3.13 and 3.14).

Energy
According to a sustainable approach, the best energy gain is the energy saving. Thus, the envelope structure consists of an innovative wood-based prefab solution with high thermal resistance. For smoothing the summer thermal peaks, the wall integrates two features: a ventilation chamber, behind the cladding layer, to reduce the heat loads of direct solar radiations, and a massive layer, to prevent internal overheating and shift the thermal flow. According to a smart energy production, the conditioning system uses an eat pump, combined with thermodynamic panels, integrating both the DHW needs and the ventilation system. Flexible Photovoltaic Panels are used as curtain wall to shade the loggia while producing energy. The total energy consumption in Contest Week accounted just 3.48 kWh/m^2 (Fig. 3.13).

Environment
RhOME is designed in order to use only recycled or natural materials and products that could reduce the energy consumptions, from the manufacturing to disposal,

Fig. 3.12 RhOME (Ph. courtesy Università di Roma Tre)

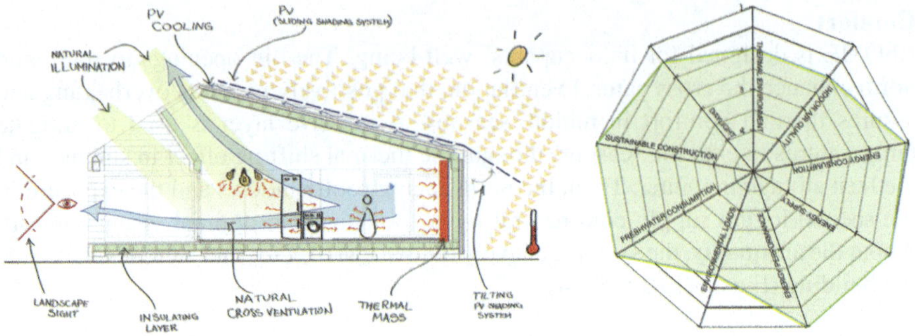

Fig. 3.13 RhOME schematic design and AH Radar (© University of Rome TRE, Politecnico di Milano, and Active House Alliance)

Fig. 3.14 RhOME details: on the left, the indoor space; on the right, the side view of its particular shape (Ph. courtesy Università di Roma Tre)

according to an LCA approach. The most innovative features in the environmental issue are the massive layer. The sand is taken from the resulting of planned demolitions of urban regeneration intervention process, or from the excavation site of the building itself (Fig. 3.13).

References

https://www.activehouse.info/

Brambilla A (2014) Mediterranean active house: analysis of climate and users impact on energy and indoor comfort in the new efficiency standard for sustainable buildings. Ph.D. thesis, Politecnico di Milano

Erell E, Portnov BA, Etzion Y (2003) Mapping the potential for climate-conscious design of buildings. Building Environ 38:271–281

Rubel F, Kottek M (2010) Observed and projected climate shifts 1901–2100 depicted by world maps of the Köppen-Geiger climate classification, Meteorol 19:135–141

Mimmack GM, Mason SJ, Galpin JS (2001) Choice of distance matrices in cluster analysis: defining regions. J Clim

Miller W, Buys L (2012) Anatomy of a sub-tropical positive energy home PEH (2012) Solar energy 86:231–242

Nicol JF, Pagliano L (2007) Allowing for thermal comfort in free-running buildings in the new European Standard EN15251. In: Conference on building low energy cooling and advanced ventilation technologies in the 21st century, Crete

Baker N, Standeven M (1996) Thermal comfort in free running buildings. Energy Build 23:175–182

Voss K, Sartori I, Lollin R (2012) Nearly-zero, net zero and plus energy buildings. How definitions & regulations affect the solutions. REHVA J 6(49):23–27

Sartori I, Candanedo J, Geier S, Lollini R, Garde F, Athienitis A, Pagliano L (2010) Comfort and energy efficiency recommendations for net zero energy buildings. In: International Conference on solar heating, cooling and buildings, Graz

de Graca GC, Augusto A, Lerer MM (2012) Solar powered net zero energy houses for southern Europe: feasibility study. Solar Energy 86:634–646

Chan K, Saltelli A, Tarantola S (1997) Sensitivity analysis of model output: variance based methods make the difference. Winter Simulation Conference, Atlanta

Campolongo F, Cariboni J, Saltelli A (2010) Enhancing Morris method. Joint Research Centre, Ispra

Heiselberg PK, Brohus H, Hesselholt A, Rasmussen E, Seinre E, Thomas S (2009) Application of sensitivity analysis in design of sustainable buildings. Renewable Energy 34(9):2030–2036

Morris MD (1991) Factorial sampling plans for preliminary computational experiments. Technometrics 33(2):161–174

Saltelli A, Tarantola S, Campolongo F (2000) Sensitivity analysis as an ingredient of modeling. Stat Sci 5(4):377–395

Brambilla A, Salvalai G, Tonelli C, Imperadori M (2017) Comfort analysis applied to the international standard active house: the case of RhOME, the winning prototype of Solar Decathlon 2014. J Build Eng 12:210–218

Van Hoof J, Hensen JL (2007) Quantifying the relevance of adaptive thermal comfortmodels in moderate thermal climate zones. Build Environ 42(1):156–170

IPCC (2014) Climate change synthesis report. http://www.ipcc.ch/. Accessed 18 May 2016

www.passive-on.org

Chapter 4
NZEB and Active House: A Case Study of Residential Building in Northern Italy

Abstract According to the European Directive, the Nearly Zero Energy Building standard (NZEB) represents a backbone to achieve the European CO_2 emission reduction target considering both, new and under renovation buildings. In this chapter, the Maison Verte, a low energy residential building, located in northern Italy has been presented showing the sustainable design approach, the technological design and his energy and comfort rating. The building represents a high energy efficient case study building, in which, the principles of comfort (visual and thermal), energy and environment are perfectly balanced according to the voluntary Active House certification scheme. During the design phase, the overall building performances has been deeply analyzed applying a multi-disciplinary design optimization process tools where parametric design, financial models and energy simulation tools are combined to accelerate the design phase and explore design options faster and more effectively than current conventional design process.

Keywords Nearly zero energy building · Sustainable building technology
Building information modeling · Design optioneering

4.1 The Nearly-Zero Energy Building Standard Definition

According to Eurostat, buildings account for 38.1% of energy consumption in the European Union causing 36% of the total EU Green Gas Emissions (GHE). In 2015, the households or residential sector represented 25.3% of final energy consumption in the EU. While new low-energy buildings generally need fewer than three litres of heating oil per square meter per year (30 kWh/m^2 year), older buildings consume about 15–25 l on average (from 150 to 250 kWh kWh/m^2 year). The Building stock observatory taking data from different European granted researches give a fairly detailed picture of the European building stock showing different indicators: thermal efficiency, energy consumption for space heating, hot water, cooling, lighting and the split applied between different heating and hot water systems and energy carriers.

This chapter is authored by Graziano Salvalai.

L. Feifer et al., *Active House: Smart Nearly Zero Energy Buildings*,
PoliMI SpringerBriefs, https://doi.org/10.1007/978-3-319-90814-4_4

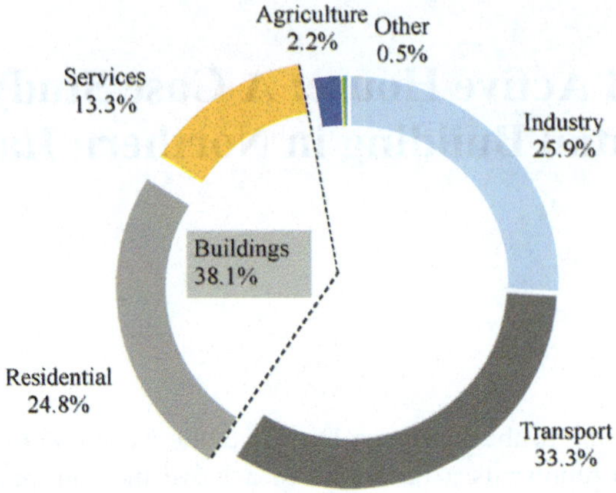

Fig. 4.1 Energy consumption by sector in the EU-28 in 2015 (Eurostat, elaborated by the author)

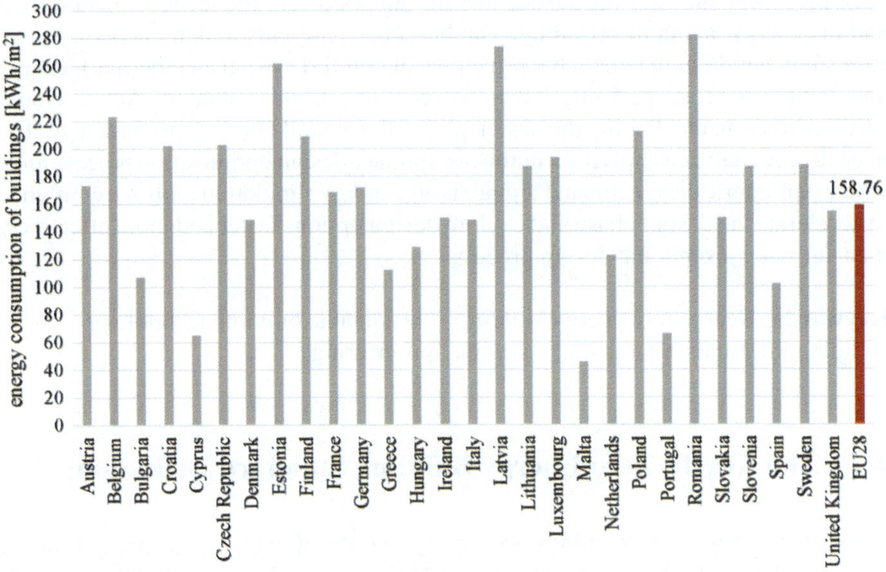

Fig. 4.2 Energy use for residential building [kWh/m^2] in different MSs (EU Building Stock Observatory database, elaborated by the author)

From this study, as shown in Fig. 4.1, the EU building stock consumes on average of 158 kWh/m^2 year, due mainly to the poor quality of the building envelope with an average level of the thermal transmission equal to 1.69 W/m^{2}K (Fig. 4.2).

Improving the energy efficiency of buildings is one of the most solid way to prepare our economy for a sustainable energy transition towards 2050 low-carbon economy target. The realistic potential for energy savings in buildings is very significant starting from now, by using energy saving materials and technologies and

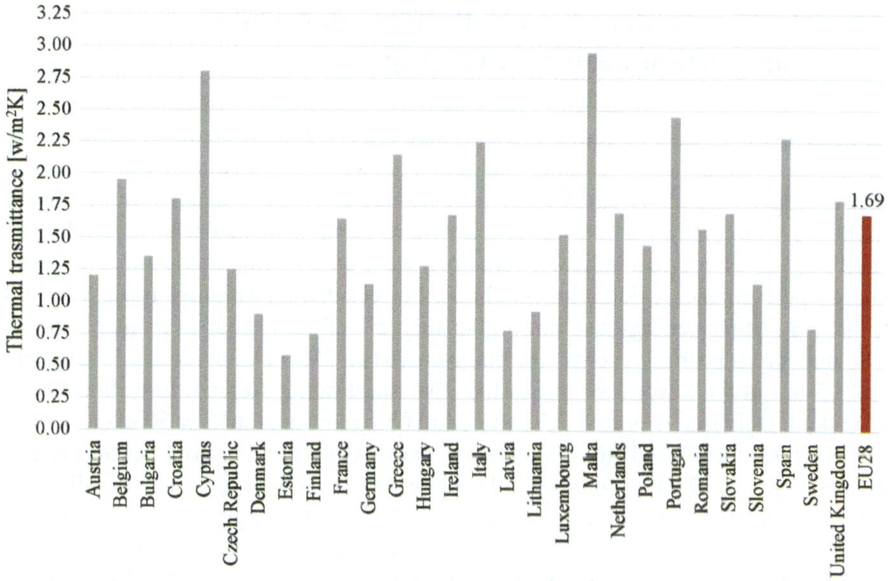

Fig. 4.3 Thermal transmittance [W/m^{2}k] of the building shell in different EMS (EU building stock observatory, elaborated by the author)

passive heating/cooling strategies that are already available on the market (Salvalai et al. 2013). Moreover by using a proper design approach; increasing the thermal insulation level; applying passive strategies and producing energy from renewable energy source, the energy demand of the EU building stock can be reduced by a fraction of current levels and can be increased the number of nZEB building across the world (Fig. 4.3).

The principle of nearly Zero Energy Buildings (nZEBs) has been defined within the EPBD Directive 2010/31/EU. The general nZEB design concept involve two main strategies:

- minimizing the need for energy use through energy-efficient measures;
- using renewable energy and applying low energy technologies to meet the remaining energy needs.

Considering the uncertainties of the renewable energy source, before producing energy is crucial to design the building, including passive heating and cooling strategies (thermal mass, shading, ventilation etc.) in order to reduce the winter/summer load. This concept starting from the idea that: every Watt that gets saved in lighting, HVAC, and plug loads is a Watt that does not need to be generated by renewable energies or sourced from the grid.

4.2 The Nearly-Zero Energy Building and Main Construction Features in Europe

Considering the nZEB building standard there is a lot of interpretation in defining the different quantitative performance indicators to be considered during the calculation method (Sesana and Salvalai 2013). Due to this reason, the European Members States (MSs) have adopted their own definition with relevant inconsistencies across Europe (BPIE 2015). Numerical indicators are not always easily comparable due to different energy performance calculation methods. For residential building most MSs aim to have a primary energy use not higher than 50 kWh/m^2 year of primary energy. The maximum primary energy use ranges between 20 in Denmark or 33 in Croatia and 95 in Latvia. Several countries like Estonia, France, Ireland, Slovakia, United Kingdom, Bulgaria, Denmark, Croatia, Malta Slovenia aim at 45 or 50 kWh/m^2 year of primary energy. In Italy the national law (D.M. 20 of June 2015) defines the primary energy limits of nZEB using a calculation process based on reference building and introduces a minimum rate of renewables. A study conducted by Paoletti et al. (2017) analyzes the construction future of a set of nZEBs collected in 17 EU countries within the EU project "Nearly Zero-Energy Building Strategy 2020" (Zebra 2020 project, 2014–2016). The project analyzes 411 nZEBs divided by new (81%) or renovated buildings (19%) and located in different EU climate zones (Fig. 4.4).

From the study, it has been highlighted that the adoption of the energy efficiency technology is not strongly affected by the climate condition. The building envelope characteristic, described by the U value (W/m^{2}k) as indicator, shows a weak correlation with the external climate conditions with strong homogeneity across different climate zones:

- Walls between 0.11 and 0.17 W/m^{2}K
- Roofs between 0.09 and 0.15 W/m^{2}K
- Floors between 0.12 and 0.23 W/m^{2}K
- Windows between 0.85 and 1.04 W/m^{2}K.

The passive design measures has been currently applied in all the analyzed case studies highlighting their relevant effect on heating/cooling load: sunshade (80%), natural ventilation (63%), night cooling (59%) and thermal mass (57%). Considering the mechanical ventilation in buildings the analysis shows a large use of this technical solution (89% of the total cases). In 84% of the cases, the ventilation system is coupled with an air to air recovery system with efficiency higher than 80%. Concerning the heating system, around 60% of the total cases use a single system for both needs (heating and domestic hot water), the most common system is the heat pump followed by boiler and district heating. The heat pump represents also the common system for cooling purpose considering the low carbon emission and the possibility to increase the load matching. As defined by the EBPD, the integration of the renewables is crucial in an nZEB[1]: the 81% of the scheduled building integrate renewable energy

[1]Most of the national nZEB definition require a minimum amount of on-site renewable energy production.

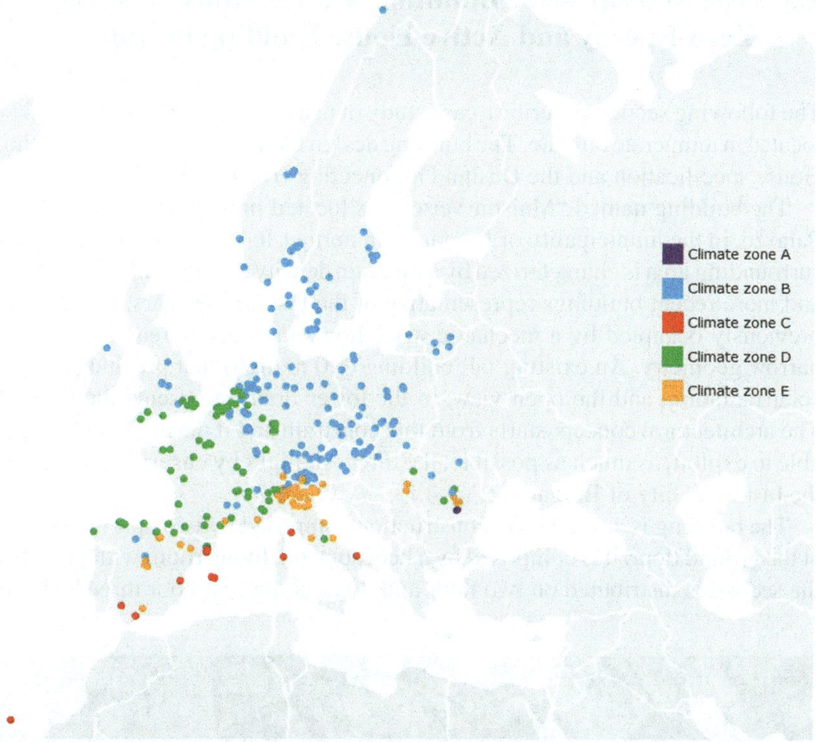

Fig. 4.4 Distribution of nZEB building according to different EU climate zones (ZEBRA project)

production technologies in the building envelope (photovoltaics and solar thermal systems). A high-efficiency HVAC system is a critical part of an nZEB, without these technologies, it would be very difficult to reach that target. Highly efficient HVAC systems use innovative technologies, such as inverters that drive the compressor and smart controls to manage the time and usage. In general, a high-efficiency HVAC system cannot be effective if it is coupled with poorly installed or designed insulation, or other building components; therefore an holistic view is needed. The Active House standard (Active House Alliance 2013) is based on a holistic view that integrates the principle of comfort, energy and environment for more energy efficient sustainable buildings.

4.3 The Maison Verte Building. A Case Study of Nearly Zero-Energy and Active House Building in Italy

The following section describe a case study of nearly zero energy residential building located in temperate climate. The building design has been optimized using the Active House specification and the Design Optioneering (DO) approach.

The building named "Maison Verte",[2] is located in the historic district of Borgo Palazzo, in the municipality of Bergamo, in norther Italy; 50 km far from Milan. The surrounding area is characterized by a medium density district with a mix of historical and more recent buildings representative of the last 20–30 years. The building site, previously occupied by a mechanic workshop, is a fenced area characterized by a narrow geometry. An existing tall building (6.0 m high) that precludes partially the solar radiation, and the open view, to the lower floor, represents the South border. The architectural concept starts from this constrain and it develops a design solution able to exploit, as much as possible, the direct sunlight by ensuring the view towards the historical city of Bergamo (Fig. 4.5).

The building is a three level construction composed by two apartments. The first, at the ground floor, it is composed by a bedroom and living room with private garden; the second is distributed on two floor and it has at the first floor three bedrooms, the

Fig. 4.5 View of the building site (Atelier2, elaborated by the author)

[2]Design and project by Atelier2—Arch. Valentina Gallotti (Architecture and Technology), VANONCINI SpA and Politecnico di Milano—Prof. Marco Imperadori (Engineering and development). Politecnico di Milano took care only of Active House simulations and labeling.

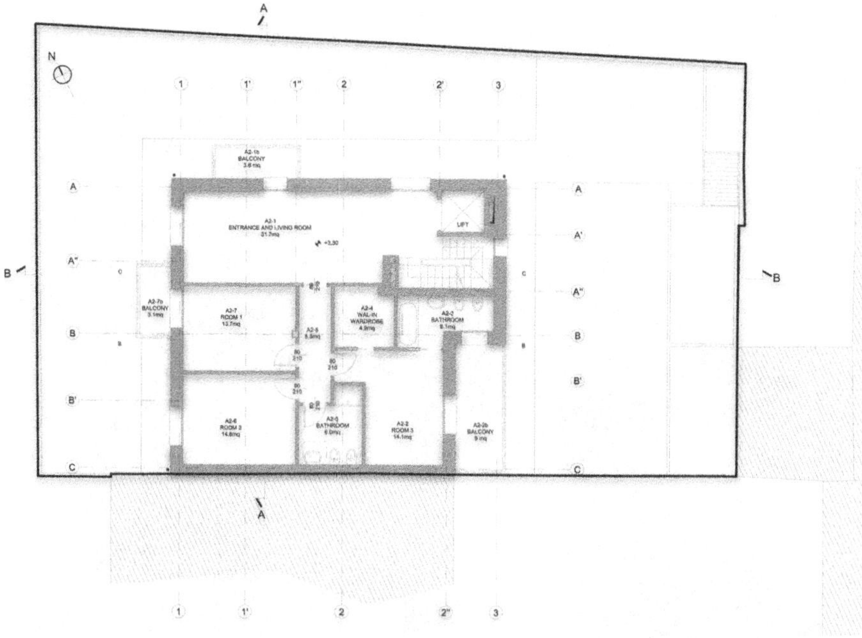

Fig. 4.6 Plan of the first floor of Maison Verte building (Atelier2)

living room and two bathroom. At the second level a large living room is faced to South with a 70 m^2 of open terrace. From the terrace, a staircase connects the floor with the green rooftop (Figs. 4.6 and 4.7).

The name "Maison Verte" refers in general to the concept of eco-sustainability and in particular, to the vegetation that characterizes the construction itself and the building site. Four levels of green have been designed: (i) the garden at the ground level giving to the lot draining garden surfaces; (ii) the vertical green of the concrete stairs block, (iii) the green of the terrace on the first floor and (iv) the green roof of the rooftop. The green roof, in addition to the mitigation effect of the high summer temperatures, allows enjoying a beautiful view of the historical city of Bergamo (Fig. 4.8).

From the technological point of view, the entire building is designed with innovative lightweight "dry-layered construction". Considering the mentioned constrains of the building lot (the South orientation, for the first two floor is obstructed by an existing building), the East and West openings have been designed by a larger dimension in order to increase the visual and the thermal comfort. The big windows faced south, located on the fourth floor, have been protected from excessive solar radiation by linear horizontal overhangs. The shading systems hold the photovoltaic solar panels for renewable energy productions. The horizontal green roof of the top hosts several skylight increasing the internal natural daylighting of the living space (Fig. 4.9).

Fig. 4.7 View of the building (Atelier2)

Several passive cooling strategies have been integrated in the project: heat gain control strategies and heat dissipation techniques allow increasing the indoor thermal comfort and reduce the total energy consumption minimizing the use of mechanical system. Figure 4.10 shows the schematic design for both, the heating and the cooling seasons. The passive solar heating involves mainly the use of thermal insulation and the collection of solar energy; the passive cooling involves, among others, the solar protection, the use of thermal mass and the natural ventilation.

Fig. 4.8 View of the historical city of Bergamo from the green roof (Atelier2)

4.3.1 Technological Design

The Maison Verte represents an example of lightweight steel structure, high-insulated building envelope with low embodied energy. The structure has been designed by "HE" profiled steel columns and "IPE" profiled steel beams that carry the vertical loads to concrete direct foundations. The horizontal loads (wind and earthquake forces) can be absorbed by diagonal "L" shaped steel profiles connected by bolts to the vertical structure (Fig. 4.11). The ground slab is aerated concrete floor composed by modular plastic formwork placed side by side creating an air layer able to convey outside moisture and Radon.

Fig. 4.9 Horizontal shading elements (Atelier2)

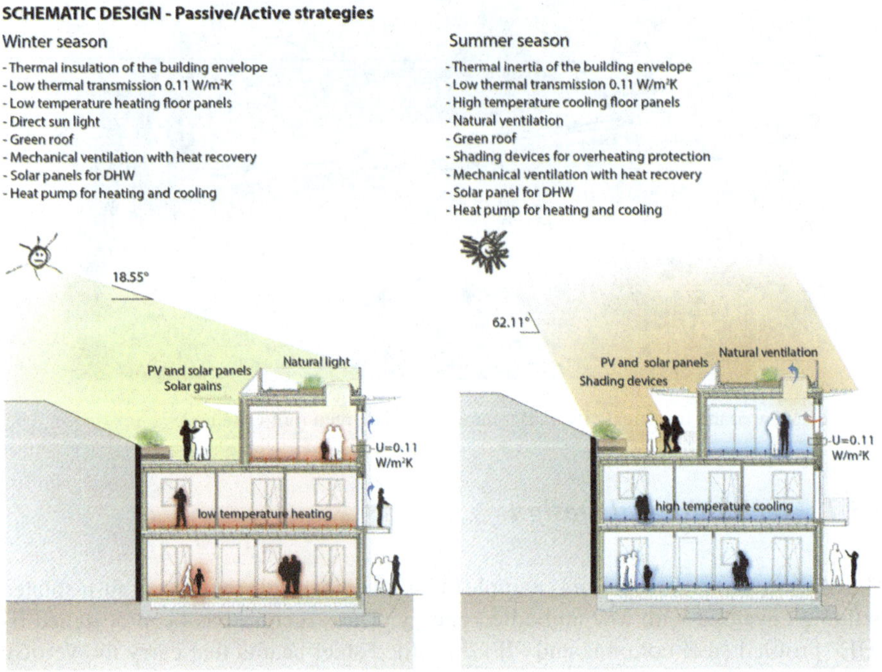

Fig. 4.10 Schematic design with heating and cooling passive/active strategies (Atelier2)

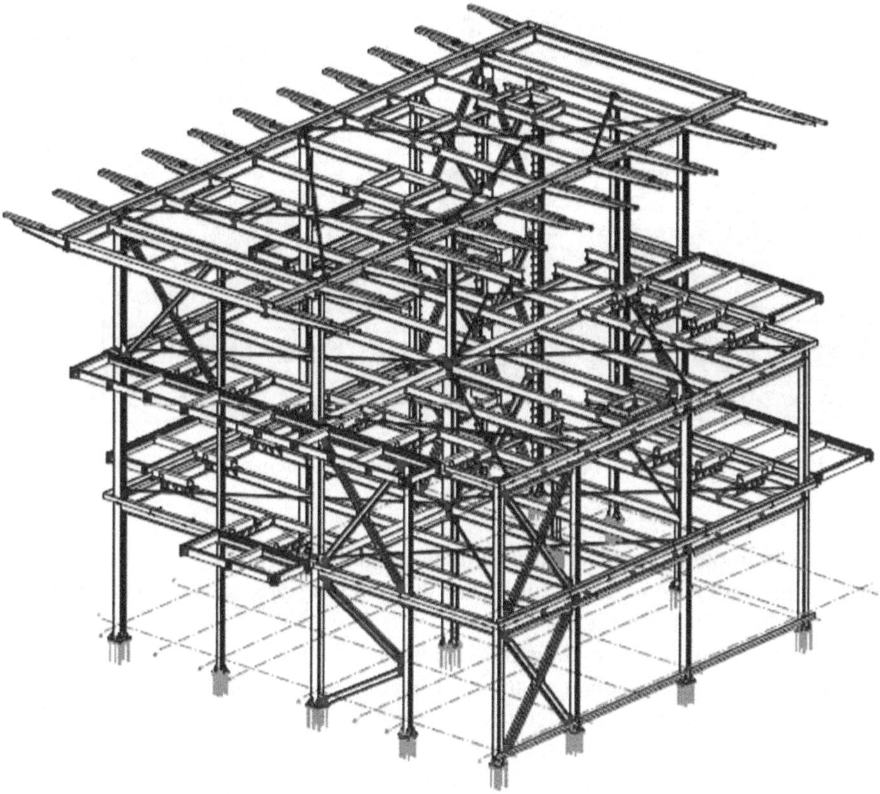

Fig. 4.11 Structural scheme of the building (Atelier2)

The horizontal slabs are composed by corrugated metal sheet below a thick layer of reinforced concrete. The building envelope is composed by several lightweight layers with high thermal resistivity and proper thermal inertia able to guaranty sufficient time lag for superior summer thermal comfort. Moreover, the different material layers increase the acoustics insulation of the building envelope (Fig. 4.12).

The vertical wall (from outer to inner inside) has been composed by thin plaster coating, expanded polystyrene panels, reinforced laminated concrete panels mounted on metal galvanized "C" shaped profiles with rigid mineral wool in separation, for a total thickness of 27.0 cm. The inner side is an internal counter-wall with metal framework, double plasterboard sheet with vapor barrier foil. The average thermal transmittance is equal to 0.11 W/m^2K. The horizontal slab (from inside to outside) is composed by suspended gypsum ceiling panels coupled with 180 mm of thermal insulation, reinforced concrete slab on steel deck and additional external insulation layer. The overall thermal transmittance is equal to 0.09 W/m^2K. The windows have been composed by white PVC frame and selective glasses with U_w equal to

Fig. 4.12 Construction phases of the structural steel elements (Atelier2, elaborated by the author)

1.3 W/m^2K, mounted on thermal insulated monoblock frame. The solar control is made with aluminum external sunscreen with arching profile positioned at 10 cm from the glazed surface (Fig. 4.13).

The building has been equipped with an air-to-water electrical heat pump connected with underfloor radiant systems for heating and cooling. During the summer season, the relative high floor temperature provide the needed cooling, while during the winter seasons the low temperature water provide the required heat. The air change rate and the control of the ambient humidity is transferred to a centralized air-handling unit with the air distribution located in the space between the deck and the suspended ceiling.

4.4 Maison Verte: The Active House Standard Evaluation

The Maison Verte is under certification according to the Active House labeling. The Active House approach was applied during the design phase as a tool for the optimization and identification of critical issues. As discusses in the Chap. 1, the Active House standard differs from others standards because it takes many more elements besides energy into consideration, it introduces a holistic approach towards building design, balancing three main principles for new and/or refurbished projects: comfort, energy, and environment. The integration of each parameter describes the level of ambition of how 'active' the building has become. For a building to be considered as an Active House, the level of ambition can be quantified into four levels where 1 is the highest level and 4 is the lowest. The next section will describe the level of ambition of the Maison Verte building showing the calculation method and the results reached for a selected set of criteria defined by the standard.

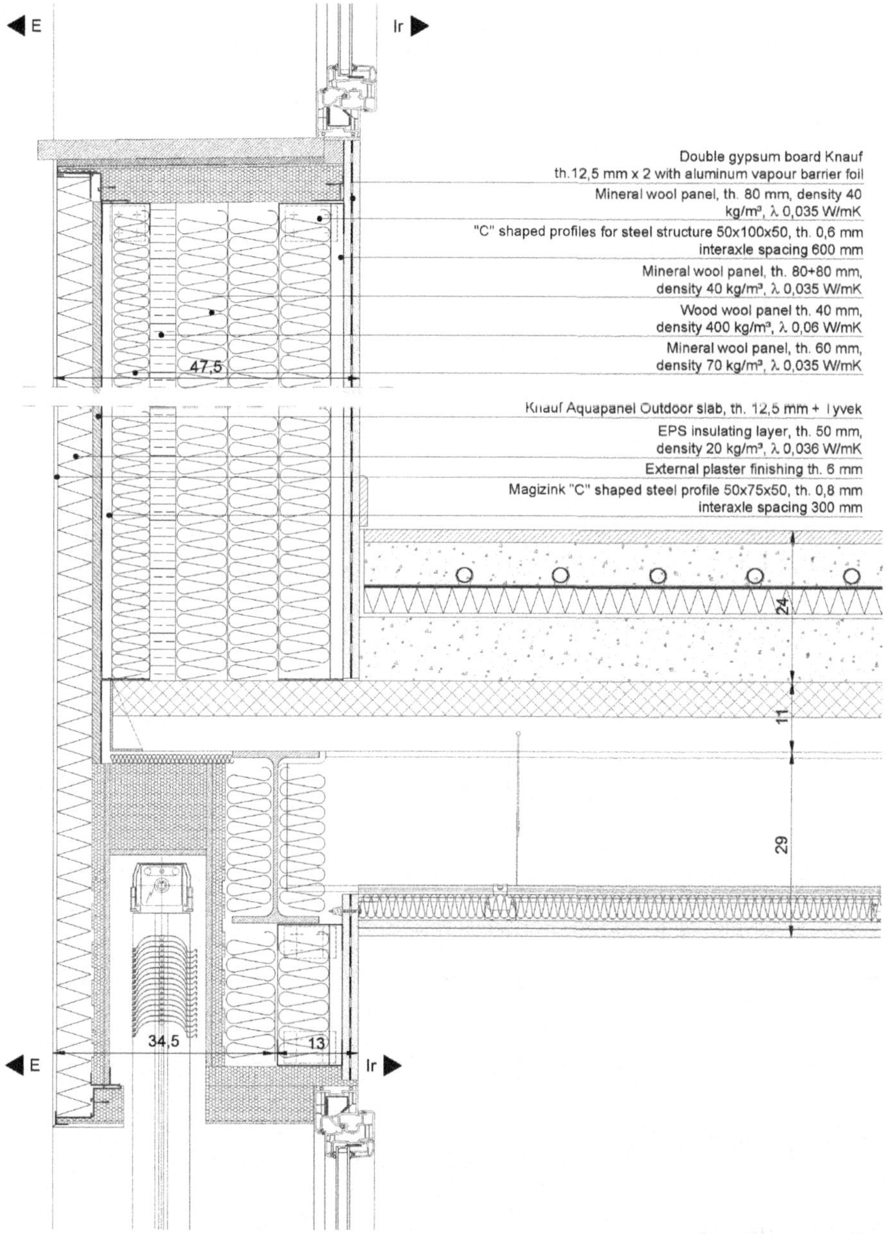

Fig. 4.13 Vertical section of the building envelope (Atelier2)

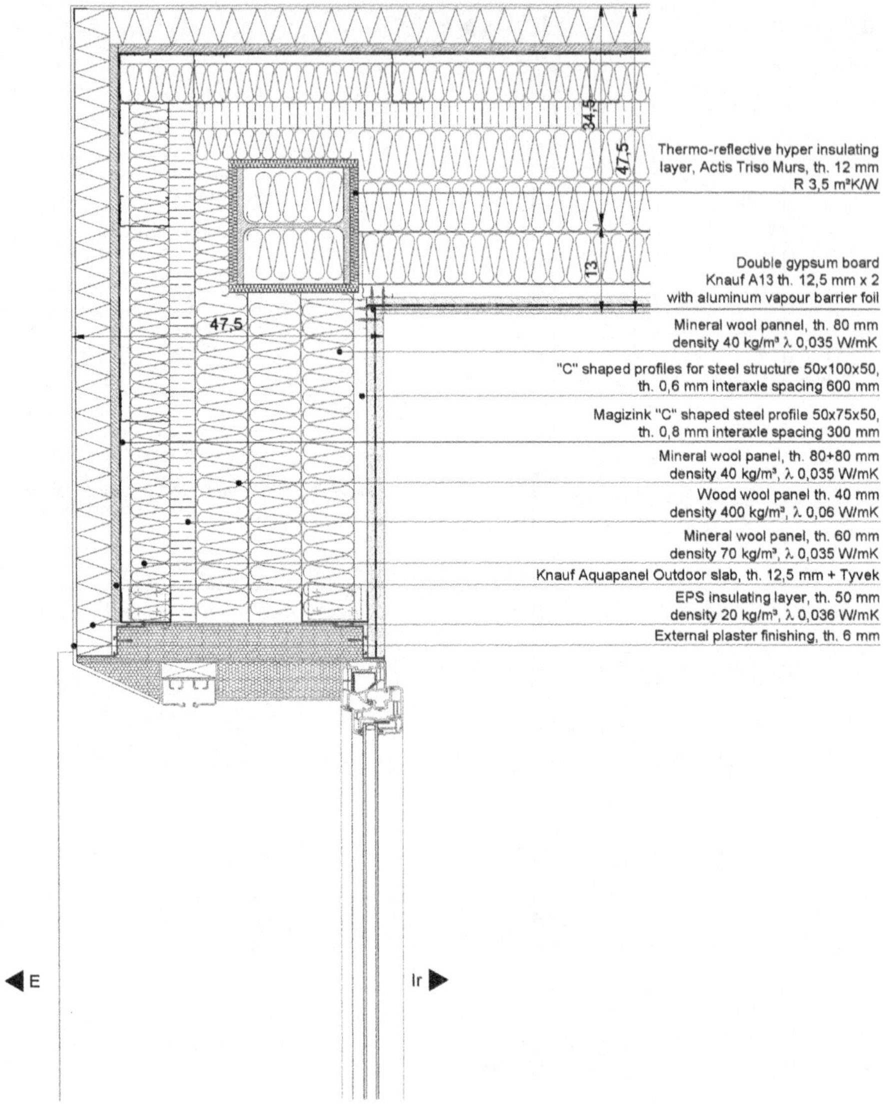

Fig. 4.13 (continued)

4.4.1 *Comfort*

People spends more than 90% of their time inside building with high impact on physical and mental health. Buildings must be designed for people, and they have to keep people comfortable, efficient, healthy, and safe according to the user expecta-

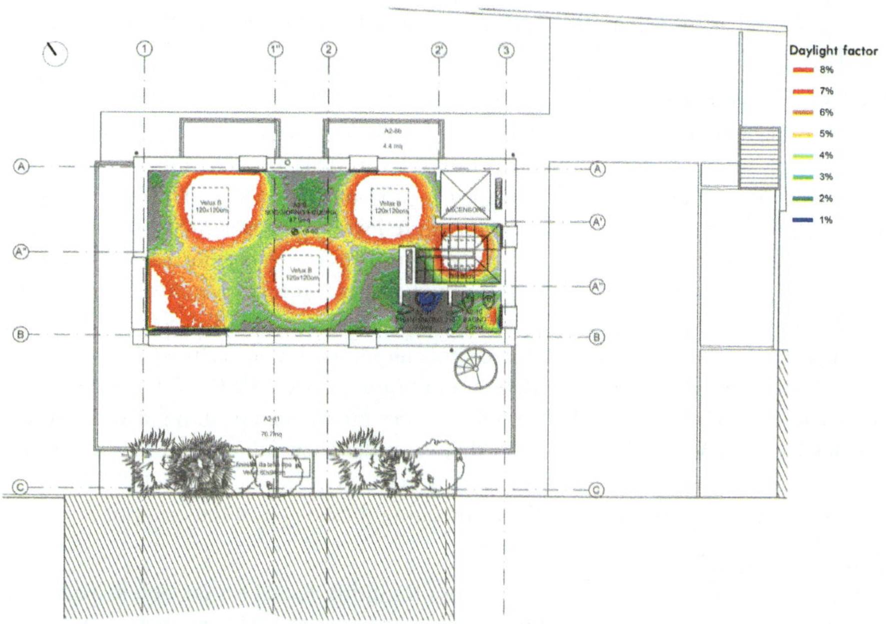

Fig. 4.14 Daylight factor analysis for the third floor (Atelier2, elaborated by the author)

tion. A house with low energy consumption and poor comfort condition is not a high quality energy efficient building. The challenge on which designer must face on is to design building with high thermal and lighting comfort involving a small amount of non-renewable energy source during the life span of the building.

Daylight

The relationship between transparent and opaque surfaces has been optimized to ensure high natural light quality and create pleasant views to the outside. From one hand big transparent surfaces improves the visual comfort and, in the other hand, a big amount of solar radiation can decrease (especially during summer season) the expected thermal sensation increasing the operative room temperature of the internal zones. Considering this dual aspect, during the design phase, was therefore necessary to carry out specific iterative evaluation controlling at the same time both the aspects. As shown in Fig. 4.14 the natural light quality level has been controlled calculating the Daylight Factor (DF) for all the internal rooms. The calculation has been performed by means of VIZ Render simulation program. Based on simulation results the mean DF values were equal to 3%, reaching the level 2 of the specific criteria set-point.

The direct sunlight availability has been also calculated. The analysis has been performed during the equinoxes according to the BS8206-2: 2008 "Lighting for buildings, Part 2: Code of practice for daylight". The simulation results shows a minimum sunlight provision higher than 10% of the probable sunlight hours. Due to this result and according to the daylight scale criteria, the Maison Verte reaches the level 1.

Thermal Environment
A good thermal comfort is a crucial aspect for comfortable and healthy home. A low energy house should be able to control overheating in summer and provide the right temperature in winter with a low amount of energy. The Active House approach request an hourly indoor temperature level analysis showing the maximum and the minimum operative room temperature (ORT) during the whole year (Salvalai 2015). The evaluation of the minimum and maximum operating temperatures has been estimated by means of dynamic simulation tool (TRNSYS v17). The summer results show, for 95% of occupied time, an internal temperature (T_i), lower than 26 °C[3]—criteria set for rooms with air conditioning system (ISO 7730:2005)—as set by the level 2 of the standard. The results were guarantee designing a well thermal insulated envelope with high energy efficient underfloor radiant cooling system coupled with an air-handling unit with air heat recovery. The minimum ORT in winter has been estimated higher than 20 °C, reaching the level 2 of the criteria.

Indoor Air Quality
Active Houses should provide good indoor humidity levels for occupied spaces and set the maximum requirement for indoor humidity. The Maison Verte ventilation strategies contemplate the use of natural and/or hybrid ventilation systems according to their best energy performance. Considering the ventilation rate, provided by the mechanical ventilation system (heat recovery efficiency higher than 80%), the indoor high quality reach high standard with an estimated CO_2 concentration level below 500 ppm (level 2) (Fig. 4.15).

4.4.2 *Energy*

An Active House is an energy efficient building equipped with systems able to exploit all the renewable energy sources available in the building side or nearby the house. The design of an Active House has to be based on the Trias Energetica[4] approach to sustainable design, focusing on the energy demand, integration of renewable energy and primary energy performance. For the specific case of Maison Verte all these

---

[3]The thermal environment has been calculated according to the EN standard 15251:2007 Indoor environmental input parameters for design and assessment of energy performance of buildings-addressing indoor air quality, thermal environment, lighting and acoustics. The set point temperature for heating and cooling is defined by the EN standard 7730:2005.

[4]The Trias Energetica, developed by the Delft University of Technology in the Netherlands in 1979, model offers a logical, three-step guide for tackling energy sustainability in the building sector: (1) first, reduce demand through energy efficiency measures and minimizing energy waste; (2) then, supply as much of the demand as possible through renewable sources such as wind, water and solar energy and (3) finally, supply any remaining demand as responsibly as possible through the efficient production and use of fossil fuels.

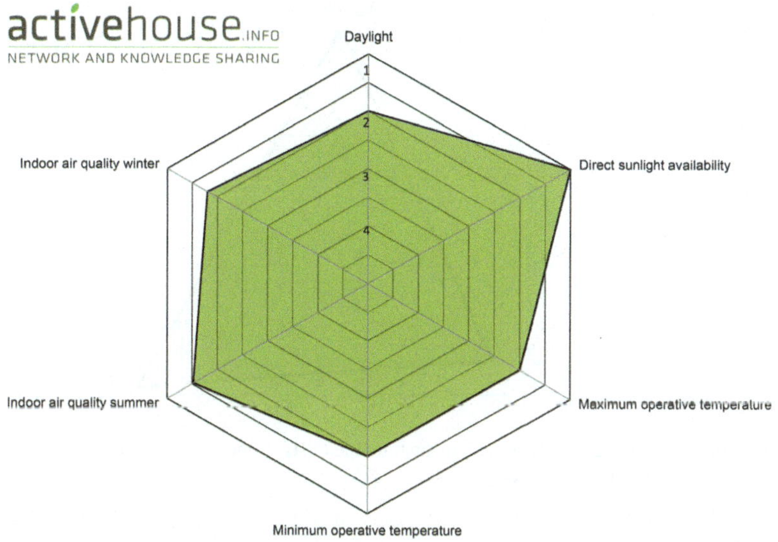

Fig. 4.15 Maison Verte comfort level: the intermediate radars shows the performance of the six comfort parameters (Atelier2)

aspects have been calculated and optimized increasing as much as possible the level of ambition.

Energy Demand
The total energy demand was calculated considering the energy for heating, domestic hot water, ventilation, cooling, control systems and lighting. According to the Active House approach the Maison Verte has been optimized with maximum use of solutions that are not energy intensive (for both the heating and cooling season): thermal insulation, solar gain control, natural ventilation, ventilative cooling etc. The building energy analysis has been conducted using the national standard (D.M. 26/06/2015) and the specific climate condition. The annual energy demand has been estimated equal to 11.5 kWh/m^2 year corresponding to the level 1 of the criteria.

Energy Supply
The energy supply from renewable sources has been calculated considering the different renewable energy technologies applied in the building: solar thermal panels, PV panels and air-to-water heat pump. The electrical energy production has been estimated using PVgis tools; the overall production considering the designed orientation, slop and solar cells typology has been estimated equal to 3500 kWh/year. The thermal energy provided by the solar thermal collectors has been estimated equal to 6800 kWh equal to 35% of the total energy need. Considering the number of user and the relative use profile the renewable energy production has been estimated to cover at least 75% of the total energy use in the building (level of ambitious equal to 1).

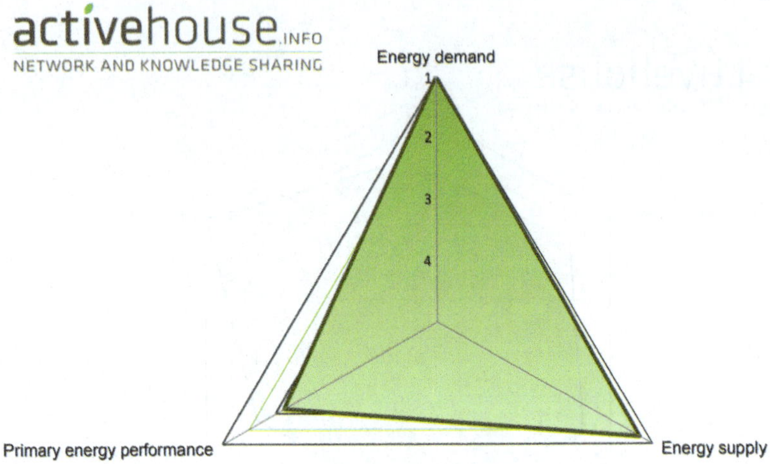

Fig. 4.16 The Trias Energetica approach to sustainable design: energy demand, energy supply and primary energy performance (Atelier2)

Primary Energy Performance

The annual primary energy performance has been estimated based on national energy factors of primary energy. The calculation includes the energy demand for the building as well as the energy supply from renewable energy, considering the proper primary energy conversion factors. The primary energy performance has been estimated equal to 13.4 kWh/m^2 year, reaching the level 2 of the criteria (Fig. 4.16).

4.4.3 Environment

To minimize the environmental impact of the Maison Verte it was necessary to evaluate and optimize the consumption of non-renewable energies and emissions into the air, soil and water caused by construction throughout its life cycle. The calculation of the Life Cycle Assessment (LCA) was carried out according to EN 15643:2011 and ISO 14040:2006. For the calculation of the impact of each specific indicator, the Inventory values (LCIA) of the German Okobau database were considered. The building's life cycle has been considered ate the following stage:

- extraction of raw materials and production of materials;
- construction process;
- operation and maintenance of the component;
- disposal at the end of service life.

The environmental impact assessment covered the following technical elements:

- roof, vertical walls, slabs, foundations and windows;

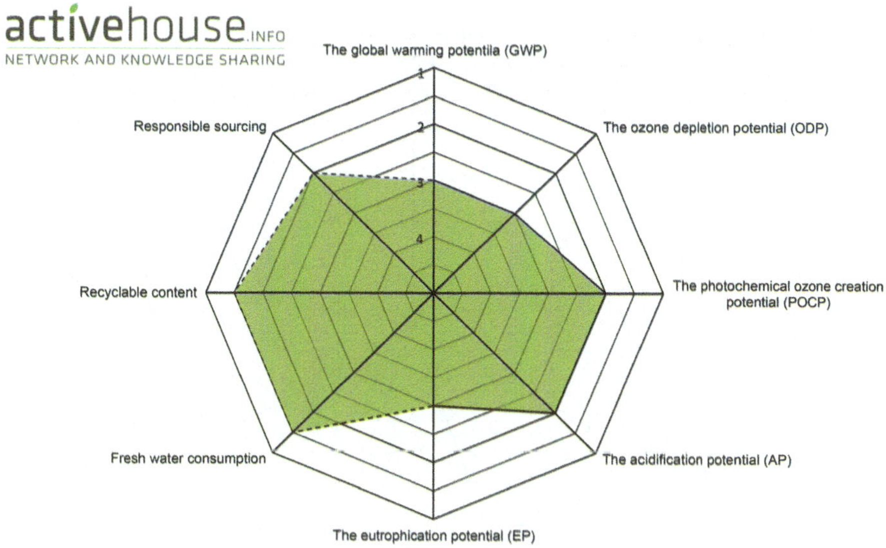

Fig. 4.17 Active House environmental indicators of Maison Verte (Atelier2)

- internal partitions, floors and ceilings;
- main technical plant components (main pipes, heat generator, plumbing fixtures, etc.).

Environmental Loads

Considering the assessment of the environmental loads, five parameters have been analyzed: global warming potential (GWP), ozone depletion potential (ODP), photochemical ozone creation potential (POCP) and eutrophication (EP). Figure 4.17 shows a comprehensive view of all the level attained for each parameters.

Global warming potential (GWP): measurement of carbon dioxide (CO_2) emission due to the greenhouse effect as results of non-renewable energy use. The parameter shows a value below 40 $kgCO_2$, eq./m^2 year, reaching the level 3.

Ozone depletion potential (ODP): measurement of the emission of gas responsible for the thinning of the ozone layer in the stratosphere. The calculated value was lower than 3.7×10^{-6} kg R11, eq./m^2 year, reaching the level 3.

Photochemical ozone creation potential (POCP): measures the ethylene emissions that govern the ozone formation potential, lower than 0.004 kgC_3H_4, eq./m^2 year. The parameter attained the level 2.

Acidification potential (AP): measures the emissions responsible for soil and water acidification. The analysis found a value below 0.075 kgS_2, eq./m^2 year corresponding to the level 2.

Eutrophication (EP): measures the impact of fertilizer emissions in the ecosystem including Nitrogen (N) and Phosphorus (P). The parameter reaches the level 3 with an absolute values below 0.0085 $kgPO_4$, eq./m^2 year.

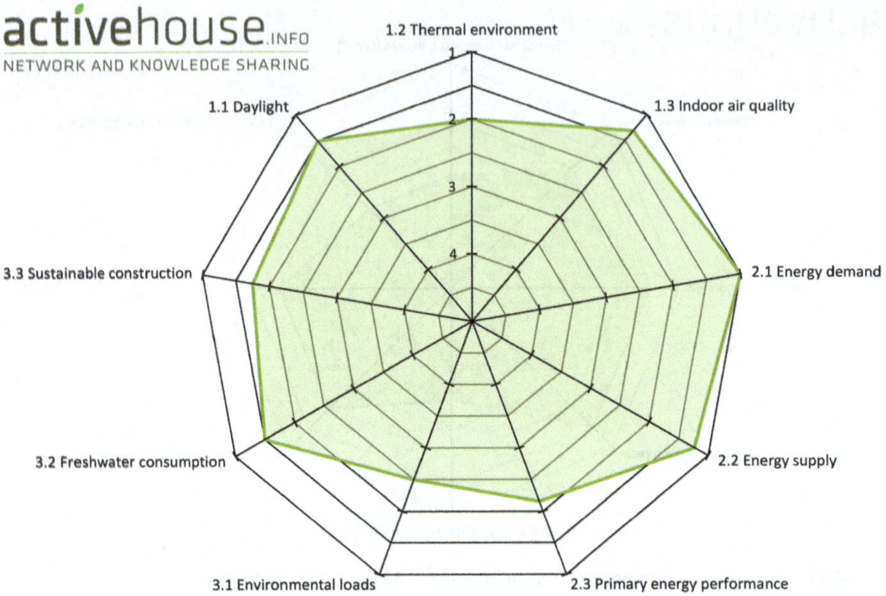

Fig. 4.18 Active House radar results (Atelier2)

The fresh water consumption, during the life-time of the building, was calculated considering different water saving strategies like the use of rain water for toilets and gardening and the use of water flow reducers. It has been estimated an improvement of 40% based on the national average water consumption (Fig. 4.17).

Figure 4.18 represents the overall quality of the building condensing all the 9 parameters evaluated. According to the Active House approach, the performance can be described through the Active House Radar showing the level of ambition of each of the three main Active House principles and their sub-parameters.

4.5 Building Information Modeling for Performance Optimization

Designer and workers involved in the design process face a challenge to deliver successful project due to several aspects that are typical of the construction industry (budget, limited information, low quality of manpower, sub-contractors, etc.). As well known the overall performance of buildings is heavily impacted by design the decisions made during the early stages of the design process. Information modeling supports in problem recognition at the initial stage, identifying the exactly location of discrepancy. Recent development regarding the use of Building Information Model (BIM) have open new scenarios increasing, potentially, the overall quality of the construction. BIM prevents errors by enabling conflict, it can provides quantities

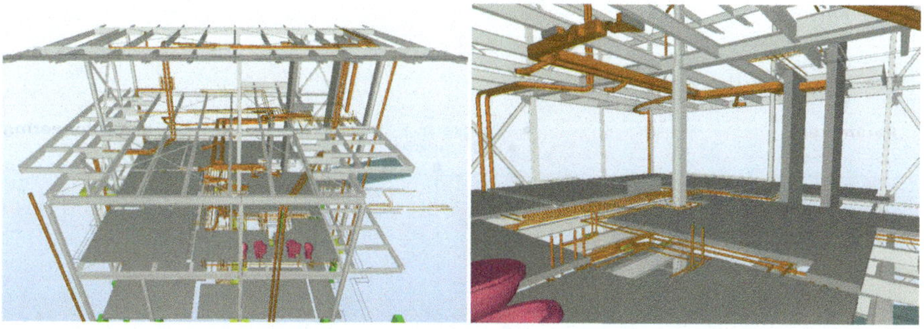

Fig. 4.19 BIM model for clash detection (Atelier2)

and properties of the materials used, can show the construction sequences to the entire facility, minimizing waste on-site and define products delivered on a just-in-time basis (Pingle and Devalkar 2017) (Fig. 4.19).

Designers are most often unable, mainly due to a limited disposition time, to explore design alternatives and their impacts on energy use or costs. The DO methodology (Gerber and Flager 2011) can be used by the design teams to explore a large number of design options more rapidly than conventional methods (Gerber et al. 2012). Using this approach, the designers can generate and evaluate different design alternatives to aid in the multi criteria analysis. The Design Optioneering involves first a range of options using associative parametric tools; then coupling the model with the integrated simulation-based analysis and, finally, using computational design optimization method to systematically look for the best option that satisfy the defined constraints (Vanossi and Imperadori 2015). The Maison Verte has been designed according to the design optioneering[5] (instead of the traditional as usual) approach involving the main actors of the construction process (owner, designers, construction company). From the methodological point of view, the approach can be divided in several phases (Fig. 4.20):

- requirement definition;
- parameters definition;
- building model;
- building simulation;
- data analysis.

The identification of the requirements is a core action since it affects the results. In the specific case energy performances, CO_2 reductions and acoustic performance were selected. The parameters represent the link between the digital model and the optimization results. A specific library, linked with technical data sheet of the different materials, has been used to populate the geometrical model.

[5]Project developed by Politecnico di Milano—Eng. Arch. Andrea Vanossi, Ph.D. and currently BIM manager of CMB Carpi, Italy.

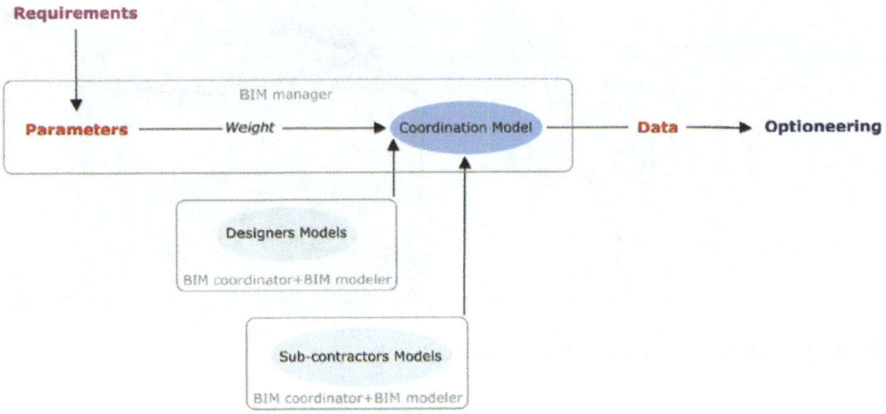

Fig. 4.20 Design Optioneering workflow applied to the case study optimization (Atelier2)

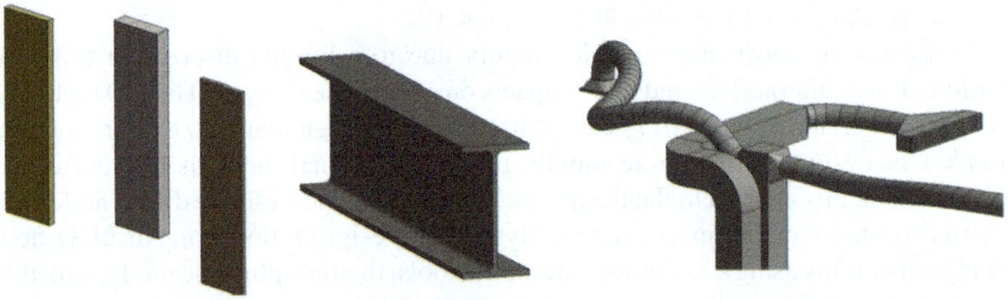

Fig. 4.21 Class of modeled building objects. From left to right: building envelope, building structure and mechanical plant system (Atelier2)

The building modeling was carried by building objects and dividing the complex structure in different discipline: the building envelope technology, the structural elements and the mechanical plant system (Fig. 4.21). This approach reduce the design time considering that the different design specialist can independently work on their specific issue and progressively merged by the BIM coordinator and identify possible critical interferences between the different objects of different disciplines (Fig. 4.22).

The last phase of the work consist on the evaluation, by multy-criteria analysis, of the different building envelope solutions taking into account the requirements and the parameters previously defined. Considering the specific case study, Fig. 4.23 shows graphically the optimal solution considering the specific case study. The building envelope chosen for Maison Verte is indeed different from what it would have been chosen using the traditional "as usual" decision criteria method that take into account only few criteria (with the construction cost as mayor constrain).

The case study underlines the potential use of the BIM during the building design phase. Particularly, the DO allows an exploitation of the parametric approach, reduc-

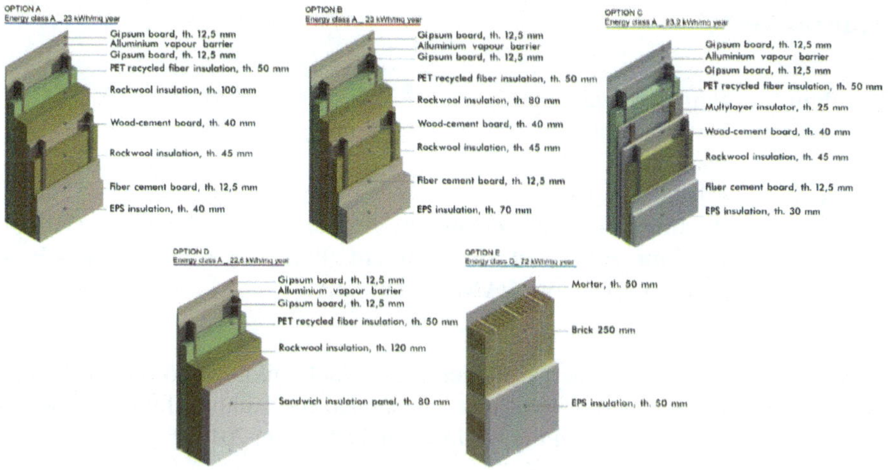

Fig. 4.22 Example of layering for cost and performance comparison (Atelier2, elaborated by the author)

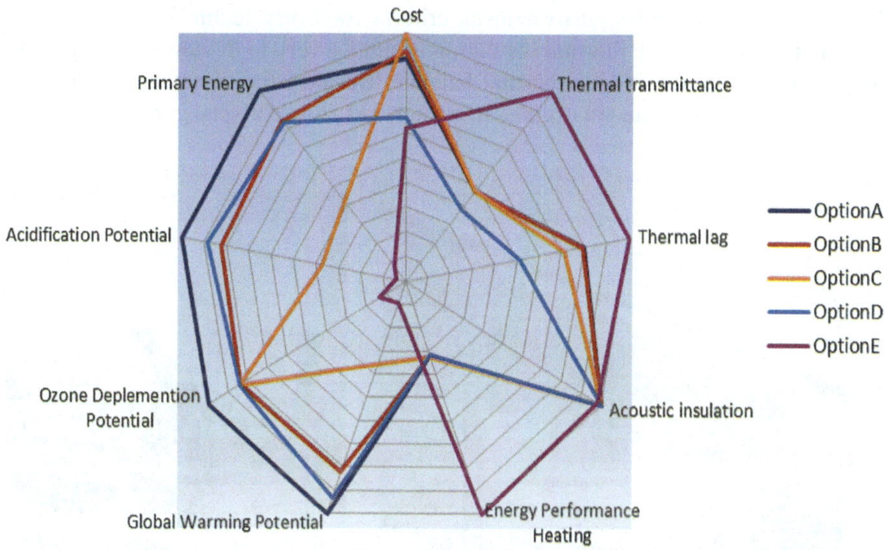

Fig. 4.23 Multy-criteria analysis and results visualization according to the Active House specifications (Atelier2)

ing the time of analysis and increasing the quality of the design process and of the overall construction.

Maison Verte

Location	Bergamo, Italy
Project type	new construction
Use	residence
Client	private company
Design	**Architecture and Technology** Atelier2—Arch. Valentina Gallotti; **Enginneering and development** Politecnico di Milano—Prof. Marco Imperadori, VANONCINI SpA
Year	2017.

Maison Verte is a high energy-efficient residential building and one of the first Italian Active Houses. The project that has defined the building follows a design optioneering approach, gaining the BIM&Digital award, in 2017 (Fig. 4.23).

Comfort
The building envelope is entirely realized through a light-weight dry construction system, where each chosen layer is the result of a design optioneering process, aiming the optimization of costs-performances criteria for every technical solution. The relevant thickness of the thermal insulation ensures a very high value of airtightness and contributes to the indoor thermal balance. Besides, the orientation and sizing of openings optimize the solar gains in winter and ensure a very good DF level. In

Fig. 4.24 Maison Verte (Ph. Atelier2—Valentina Gallotti and Marco Imperadori—Politecnico di Milano)

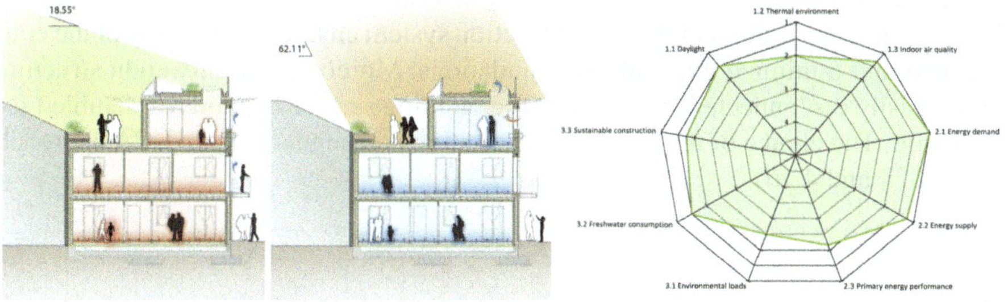

Fig. 4.25 Maison Verte schematic design and AH Radar—under validation (Politecnico di Milano and AH Alliance)

Fig. 4.26 Maison Verte details about the dried technological solutions of the envelope (Ph. Atelier2—Valentina Gallotti and Marco Imperadori—Politecnico di Milano)

the flat roof, 5 large roof windows provide a distributed brightness which increases significantly the lighting quality of the attic. The indoor panels of ceilings and walls help to keep the air clean and healthy, while the windows opening activates the natural ventilation, refreshing the air (Figs. 4.24 and 4.25).

Energy

The envelope design adopts innovative and energy-efficient technical solutions, based on dry construction systems. An additional multi-layered reflective insulation produced by Actis was installed internally in order to increase the capacity to store the energy in winter. A wide flat green roof covers the building, providing a massive shelter able to increase the thermal capacity of the envelope and decrease the temperature picks during the hottest months. A 3 kWp photovoltaic field will be installed on the flat roof, providing also an effective shading system during the summer. Moreover, four evacuated tube solar collectors for the production of hot water will the heat for service use (Fig. 4.24).

Environment
The adoption of a lightweight dry construction system ensures the rapidity of the construction and minimize the loads on foundations. Moreover, a lightweight structure can be easily disconnected at the end of the building service life and reassembled for a new life. The choice of the materials was defined through a holistic LCA approach (Figs. 4.24 and 4.26).

References

Eurostat database of the statistical office of the European Union. Available at: http://ec.europa.eu/eurostat/data/database

EU Building Stock Observatory. Database available at: https://ec.europa.eu/energy/en/eubuildings

Salvalai G, Pfafferott J, Sesana M (2013) M., Assessing energy and thermal comfort of different low-energy cooling concepts for non-residential buildings. Energy Convers Manag 76:332–341

Sesana MM, Salvalai G (2013) Overview on life cycle methodologies and economic feasibility for nZEBs. Build Environ 67:211–216

Buildings Performance Institute Europe (BPIE), Nearly zero energy buildings definitions across Europe (2015). The nZEB factsheet is available at: www.bpie.eu/nzeb-factsheet.html

Paoletti G, Pascuas RP, Pernetti R, Lollini R (2017) Nearly zero energy buildings: an overview of the main construction features across Europe. Buildings. https://doi.org/10.3390/buildings7020043

Nearly Zero-Energy Building Strategy 2020 EU founded project, 2014–2016. Project results available at: http://zebra2020.eu/resources/project-publications/

Active House Alliance. Website at: www.activehouse.info

Velux daylight Visualizer. Available at: www.velux.it/professionisti/velux-daylight-visualizer

BS 8206-2:2008. Lighting for buildings. Code of practice for daylighting

TRNSYS software. http://sel.me.wisc.edu/trnsys/features/features.html

ISO 7730:2005. Ergonomics of the thermal environment—Analytical determination and interpretation of thermal comfort using calculation of the PMV and PPD indices and local thermal comfort criteria

Decreto Ministeriale 26 giugno 2015. "Adeguamento del decreto del Ministro dello sviluppo economico, 26 giugno 2009 - Linee guida nazionali per la certificazione energetica degli edifici"

Photovoltaic Geographical Information System (PVGIS) tool. Available at: http://re.jrc.ec.europa.eu/pvgis/

EN 15643-2:2011 Sustainability of construction works—Assessment of buildings—Part 2: Framework for the assessment of environmental performance

ISO 14040:2006 Environmental management—Life cycle assessment—Principles and framework

Online tool for Life Cycle Assessments (LCA), available at: http://www.oekobaudat.de/en/info/research-projects.html

Pingle DD, Devalkar RV (2017) BIM for design optioneering: an approach to future-proofing commercial buildings. Int J Res Appl Sci Eng Technol 5(II)

Gerber DJ, Flager F (2011) Teaching design optioneering: a method for multidisciplinary design optimization. Comput Civil Eng 883–890

Gerber DJ, Lin S, Pan B, Solmaz AS (2012) Design optioneering multi-disciplinary design optimization through parameterization, domain integration and automation of a genetic algorithm

Vanossi A, Imperadori M (2015) BIM e optioneering in edifici S/R di piccola scala BIM and optioneering in dry technology small-scale building. https://doi.org/10.13140/rg.2.1.2353.0400

Salvalai G (2015) 2020. Edifici ad energia quasi zero (nZEB). Maggioli Editore

Chapter 5
Relevant Case Studies: A Benchmark for Future Design

Abstract Active House Alliance proposes a new tool to help designers in the whole construction process, since the architectural concept, through the definition of energetic strategies, until the characterization of material and construction details. The AH Radar guides the design process and, at least, evaluates the project, according to the holistic integration of its three principles: Comfort, Energy, and Environment. This final chapter collects the evaluation of 16 Active House projects, analyzing the design strategies and evaluating them through the criteria of the AH Radar. All those case studies represent real virtuous examples of good design practice, available now.

Keywords Active house · Case studies · Buildings evaluation · AH radar
Weighted criteria

5.1 Active House: Beyond 2020, Now!

The EPBD Directive 2010/31/EU (2010) stated that all the new buildings will have to reach a close-to-zero energetic annual balance between produced and absorbed energy, by the end of 2020. Buildings, indeed, are still nowadays heavily energy consumers, accounting for 40% of overall energy consumption, just in Europe. This depicts an urgent situation, where the passive behavior is no more tolerable. In addition, there are no globally shared definitions of Nearly Zero Energy Building, since each European state could define and detail its own calculation method.

In this uncertain and vague scenario, the Active House Alliance is currently into the dissemination of a more sustainable and energy-saving policy for the design and management of new and existing buildings: the Active House Vision, further deepened into the Active House Guidelines and Specifications (www.activehouse.info).

This chapter is authored by Federica Brunone.
All projects data and information had been provided by Active House Alliance, VELUX Group, and designers, and then edited by the author.

L. Feifer et al., *Active House: Smart Nearly Zero Energy Buildings*,
PoliMI SpringerBriefs, https://doi.org/10.1007/978-3-319-90814-4_5

Therefore, the AH Alliance is directly active on the topic, especially by promoting the application of the Active House principles to real projects and existing buildings. This practical approach straightly proves the feasibility and reliability of these policies for current, and future, design practices.

5.1.1 *The Active House Label and Award*

Besides the continuous dissemination and the upgrading of the Active House Guidelines and Specifications, other two communicative instruments help the diffusion of the Active House vision: the labeling and the competition.

The Active House label is a worldwide stamp for buildings, to prove their quality in comfort and sustainability. It could be issued to projects which have been designed through a combined focus on comfort, smart use of energy and minimum environmental impact, and which have been evaluated, according to the Active House directives, achieving the optimization of those three principles.

In order to involve always more intensively professional house builders, developers, architects, designers, and construction experts, every year the Active House Alliance convenes a global competition of projects about comfortable and sustainable buildings: the Active House Award. A jury of Active House experts evaluate the submitted projects from all over the world, according to the Active House vision, and the winner is awarded the Active House Label (Fig. 5.1).

5.1.2 *AH Relevant Case Studies*

The collected case studies have been selected across all over the world, to show how the AH design principles represent an affordable challenge for everyone, everywhere (Fig. 5.2). Table 5.1 lists all the analyzed projects, that are presented in accordance with the relevant realization date and distinguishing the refurbishment from the new constructions.

Fig. 5.1 Active House Award winner projects: **a** Solhuset, 2016; **b** Green Solution House, 2016; **c** Isobo Aktiv, 2016; **d** House on Garda Lake, 2016; **e** Reborn Home, 2017; **f** RenovActive, 2017; **g** Optima House, 2017. *Credit* Active House Alliance

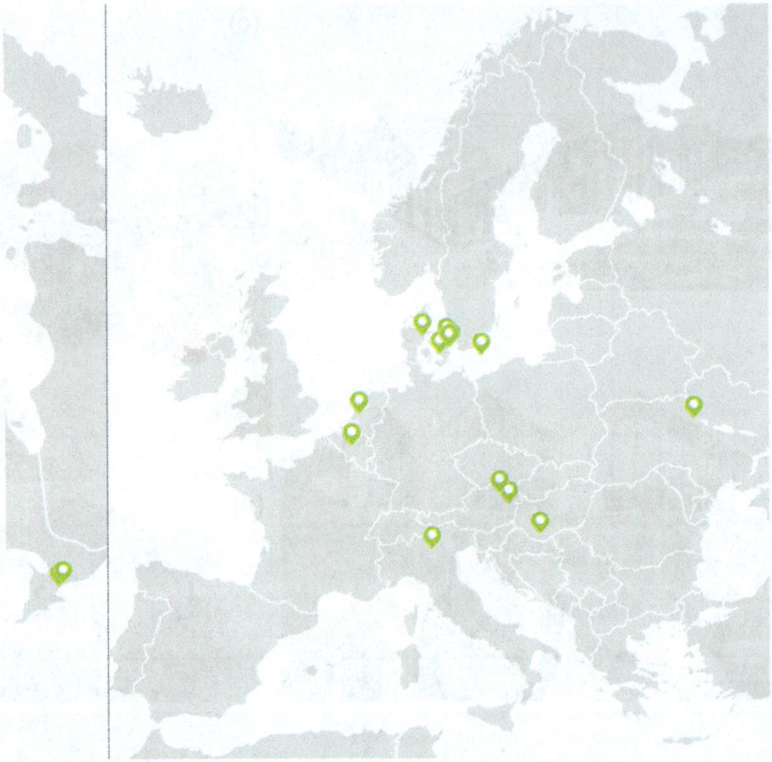

Fig. 5.2 Active House case studies are spread up all over the world. Their diffusion started from Northern Europe, and then involved the Mediterranean regions, till going overseas to North America (Canada), proving the feasibility and global efficiency of the Active House vision; here, an Active House map is presented

Table 5.1 Active House relevant case studies list

S.No	Project	Location	Year	Project type
1	Renovactive	Belgium	2017	Refurbishment
2	Reborn Home	Hungary	2017	Refurbishment
3	Copenhagen International School	Denmark	2017	New construction
4	Active House Erasmushove	Netherlands	2017	New construction
5	Great Gulf AH Centennial Park	Canada	2016	New construction
6	Green Solution House	Denmark	2015	Refurbishment
7	Optima House	Ukraine	2015	New construction
8	House by the Garda Lake	Italy	2014	New construction
9	House by the Garden of Venus	Austria	2013	Refurbishment
10	Great Gulf Active House	Canada	2013	New construction
11	The Poorters van Monfort	Netherlands	2013	Refurbishment
12	House on the Moor	Austria	2012	New construction
13	Isobo Aktiv	Norway	2011	New construction
14	Sunlight House	Austria	2010	New construction
15	Green Lighthouse	Denmark	2009	New construction
16	Home for Life	Denmark	2009	New construction

Case Study

RenovActive

Location	Anderlecht, Belgium
Project type	refurbishment
Use	social housing complex
Client	private company
Design	Antwerpbased architectural firm ONO architectuur, Le Foyer Anderlechtois and VELUX Group
Year	2017.

RenovActive is more than a building project. It represents a design concept, based on seven strategies, with the aim to achieve the balanced integration of Active House principles with the tight budget framework of a social house, proving the financial viability of Active House renovation. Modularity, flexibility, and scalability of solutions make the prototype become a stereotype, a template for 86 similar renovation projects in the community (Fig. 5.3).

Comfort
The renovation starts with the aim to improve the indoor visual and thermal comfort, assuring a pleasant and healthy environment. New, efficient, roof and façade windows, with dynamic solar blinds, help to control excess heat and to reach excellent performances on DF level. They provide also passive solar gains in winter and fresh air in summer, thanks to an automatic window-opening system. It involves the entire dwelling, because of the stack effect through the new stairwell—a respiratory channel. Moreover, temperature, humidity, and CO_2 sensors guarantee optimal indoor air quality, as the ventilation unit control adapts indoor levels to tenants' needs, in real time, saving energy and avoiding indoor discomfort (Figs. 5.4 and 5.5).

Energy
The design process focuses on the evaluation of different scenarios, combining different technical solutions and evaluating them according to the AH Radar and financial sustainability. As the social housing budget does not fit to the high costs (and low reproducibility) of a heat pump and solar collectors, designers decide for an efficient gas condensation boiler, connected to the existing network. Besides, the new insulating layer of the envelope and the floor heating system helps to provide indoor thermal comfort, while reducing the total energy consumption (Fig. 5.4).

Environment
According to the environmental sustainability required by AH standards, a 50,00l tank for collecting and recycling rainwater satisfies the flushing toilets, washing machine, and conservation of green spaces needs. Finally, the design concept upgrades to the neighborhood overall refurbishment. It will mean the establishment of a community energy system that could cover the energy needs, through renewable sources energy network (wind turbine, solar and PV panels…) (Fig. 5.4).

Fig. 5.3 RenovActive (Ph. Adam Mørk)

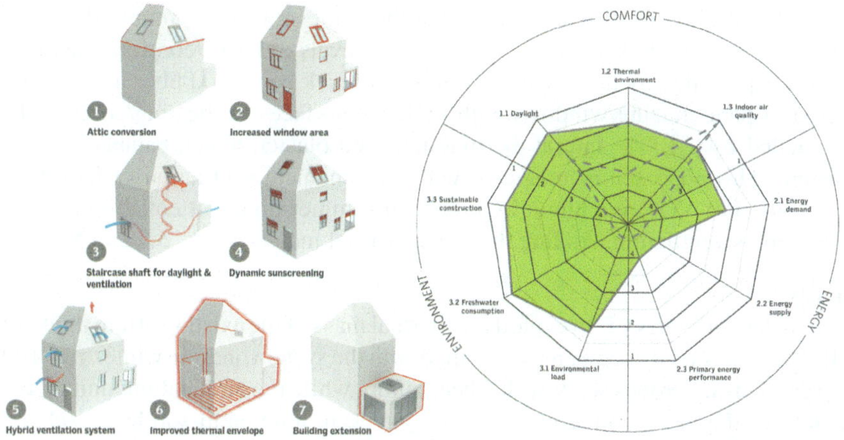

Fig. 5.4 RenovActive schematic design and AH Radar (© VELUX Group and AH Alliance)

Fig. 5.5 RenovActive: on the left, the new bright attic; besides, a night view (Ph. Adam Mørk)

Reborn Home

Location	Pecs, Hungary
Project type	refurbishment
Use	family house
Client	private
Design and development	Prof., Ph.D., Dr. Habil István Kisteledgi Dla, Kátái Nóra
Year	2017.

New functions and spaces for the refurbishment of an existing house in Hungary. The design strategies exploit the existing building features integrating them with new designing choices about the envelope and systems (Fig. 5.6).

Comfort
The new envelope includes a south-oriented solar chimney, with a large glass curtain wall and glass roof, which guarantee the achievement of optimal daylighting conditions, even in darker spaces. Besides high levels of visual comfort, the thermal comfort is provided by the integration of 36 cm thick eco-wood fiber insulation and external 3-pane glazing windows with radiative surfaces for heating/cooling. They are cladded by a 100% natural thermal-activated plaster, which balances heat and moisture. This approach is integrated with a mechanical ventilation system, for the optimization of VOC and humidity levels. In summer nights, it collaborates with a natural cross ventilation, cooling the inner thermal masses (Figs. 5.7 and 5.8).

Energy
The first design strategy is to exploit the thermal mass of the existing 30 cm thick brick wall. The envelope strategies are combined with the system ones. In winter, two 100 m deep geothermal probes work with a heat pump, while the air handling unit recovers the 90% of the heat/cooling production, by a 30 m long earth-soil heat exchanger. In summer, a passive cooling strategy is applied through natural cross ventilation. A solar thermal collector provides 100% DHW. Moreover, 10,001 hot-water and 8751 cool-water tanks ensure high efficiency in heating/cooling, by storing energy. A PV-system is located on the new roof, whose 27° slope assures a maximum yield of solar electricity by 3.5 kWp. Thus, the primary energy consumption in winter is 22.82 kWh/m^2 year, saving up to 70%. Finally, a smart management system controls heating/cooling, mechanical ventilation, and DHW (Fig. 5.7).

Environment
In order to save the existing construction, almost 100% of the load-bearing structure has been preserved, with a positive influence on LCA validation. The existing reinforced concrete slabs and brick-walls are integrated by 100% natural adobe plaster, and wood fiber insulation. Metal roofing and insulation slabs replace existing toxic materials. A 27 m^3 cistern collects water from the roof, which is recycled for washing and cooking, by using a reversed osmosis water filter (Fig. 5.7).

Fig. 5.6 Reborn Home Street front view of the dwelling with entrance and garage (Ph. The Greypixel Workshop)

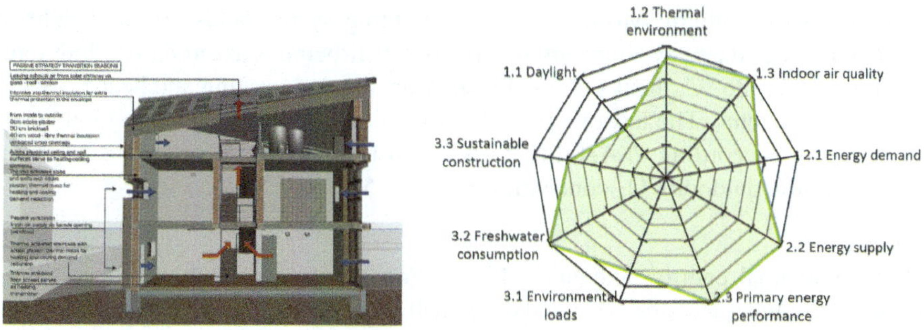

Fig. 5.7 Reborn Home schematic design and AHRadar (© Kistelegdi and Active House Alliance)

Fig. 5.8 Reborn Home solar chimney for passive ventilation, night cooling in summer and for high visual comfort levels all over the year (Ph. The Greypixel Workshop)

Copenhagen International School

Location	Copenhagen, Denmark
Project type	new project
Use	educational building
Client	private foundation
Design	C.F.Møller Architects
Year	2017.

Copenhagen International School (CIS) is a private educational institute, whose cladding system represents the largest European BIPV—Building Integrated Photovoltaic. Its installation takes the building into the Danish ForskVE/EUPD project, based on the promotion of renewable sources and high energy-efficiency (Fig. 5.9).

Comfort

Several international studies attest that natural light is essential for educational activities because it helps to improve focus and learning abilities. Therefore, the CIS' façades present an average ratio of 20% windows area over floor area, providing good levels of daylight. Moreover, a LED lighting system adds artificial light, if needed. Besides the visual comfort, the thermal well-being is assured by indoor temperatures of 21–26 °C, thanks to good insulation, heating/cooling and heat-recovery ventilation systems. The last one provides high air exchange, achieving good levels of air-quality, according to AH directives: almost all the rooms present $\leq$350 ppm levels over the outdoor CO_2 concentration (Figs. 5.10 and 5.11).

Energy

The overall final energy demand is 15 kWh/m^2y, representing a virtuous sample among the standard average demand of Danish school buildings. The well- insulated envelope—U-values of 0.74, 0.1, 0.11 and 0.12 W/m^2K, respectively of windows, external walls, roof and ground floor—defines 10.4 kWh/m^2 as the actual heating use, covered by the Central district system. Besides, the electricity supply for yearly operation (13.4 kWh/m^2) is guaranteed by both the local grid and the 6000 m^2 BIPV system (annual production = 10.7 kWh/m^2). Its modules are 60 W green chromatic coated hardened glass panels (700 $\times$ 716 mm), tested in a wind tunnel to certify the noise reduction. Considering all the contributions, renewable energy sources supply 69% energy demand, with a final energy consumption of 6.7 kWh/m^2 year (Fig. 5.10).

Environment

The design strategies of CIS lead it to become an example of a Nearly Zero Energy Building, whose definition declares high energy-efficiency and renewable sources use, with a consequent lower environmental impact. In addition to the BIPV system, the district heating and cooling systems of Copenhagen and the Danish electrical grid—almost covered by green energy sources (respectively 58% and 50%)—supply the energy demand of the building, helping to lower the CO_2 emissions.

Fig. 5.9 Copenhagen International School (Ph. Adam Moerk)

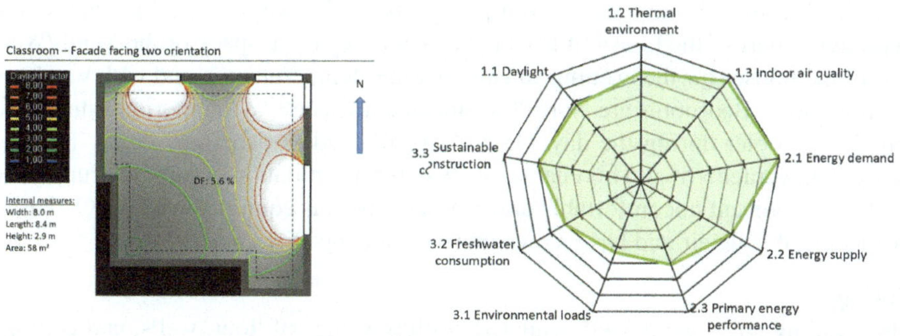

Fig. 5.10 Copenhagen International School: FLD values for a type-classroom and AH Radar—under validation (© KUBEN Management and AH Alliance)

Fig. 5.11 CIS: on the left, the new bright attic; on the right, a panoramic view (retrieved from https://www.cis.dk/); in the center, the pitched PV panels (Ph. Peder Vejsig Pedersen)

Active House Erasmushove

Location	Den Haag, Netherlands
Project type	new construction
Use	family house
Client	private end user
Design	Klimaatbouw
Year	2017.

Active House Erasmushove is an Active House since the very first design sketches. It has an educative purpose to act as an example for architects, consultants, and end-users: more than 150 people visited the building site to learn about its design principles and technical solutions, during the construction phase (Fig. 5.12).

Comfort
Comfort was the leading design principle, with both ample daylight and ventilation grills from at least two directions in every main room. The layout of the house is optimized to benefit maximally from passive natural resources. The kitchen is the main living part of the house. It is a large, two-storey open space of the South/West part of the house (for the evening sun and warmer temperatures), and with windows on three sides. Bedrooms are North/East located, to access to the morning sun while remaining cooler during the day. Daylight is allowed to penetrate deeply into the house (DFaverage > 5%). This enables stack natural ventilation in summer, supplying fresh air to prevent uncomfortable drafts. Every room is equipped with a CO_2-based extraction of stale air and has its own thermostat (Figs. 5.13 and 5.14).

Energy
The building envelope is well insulated, with R-values of floor, walls, and ceilings respectively at 5.0, 6.5 and 8.0 m^2K/W and triple glazed surfaces. Location and sizing of windows are optimized for maximum passive solar gains in winter and minimal overheating in summer. Heat comes from stale ventilation air through an air/water heat pump. It is helped by outdoor air, extracted from the cavity underneath the integrated PV panels, on roof. The heating system exploits a water-based capillary floor, walls and/or ceilings (depending on room), for a low-temperature radiative heating. A smart home system monitors all energy flows, whose preliminary results suggest a positive energy profile for 10 months per year (Fig. 5.13).

Environment
The main load-bearing construction comes from cellular concrete, which has 1/4 of the standard concrete environmental loads. Inner walls are plastered with adobe; outer wall cladding consists of douglas timber, treated with biobased Scandinavian paint. The windows and glass are Cradle to Cradle certified. Horizontal roof surfaces feature a green roof to catch rainwater and improve biodiversity. Freshwater consumptions are less than half of the average in The Netherlands.

Fig. 5.12 Active House Erasmushove (Ph. Bas Hasselaar)

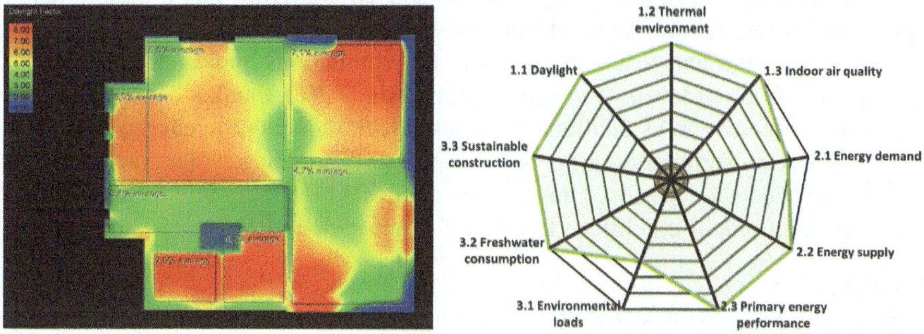

Fig. 5.13 Active House Erasmushove daylight factor simulation, via VIZ Daylight Visualizer and AH Radar (© Bas Hasselaar and AH Alliance)

Fig. 5.14 Active House Erasmushove: on the left, the open space kitchen and living room; on the right, the southern façade, the green roof and the PV-integrated roof (Ph. Bas Hasselaar)

Great Gulf Active House Centennial Park

Location	Etobicoke, Toronto, Ontario, Canada
Project type	new construction
Use	single detached residential house
Client	private user
Design	**Architectural and technological design project** Superkul, HomeCAD, HVAC Designs, Building Knowledge, Quail Engineering, Brockport (Home Technology), Velux Denmark, Great Gulf
Year	2016.

Great Gulf AH Centennial Park is the first certified Active House in the world. It is beyond the current energy performance criteria, of the Ontario Building Code 2012. Design simulations and a monitoring campaign enhance the feasibility of a new low-energy and environmental-friendly way of healthy living (Fig. 5.15).

Comfort
Great Gulf AH Centennial Park is an example of qualitative and quantitative optimization of technology and design. The C-shaped courtyard has a double-height open plan, where natural light and breeze overwhelm the indoor space, providing great spaciousness and visual connection. DF simulations and as-built monitored data confirm optimal levels of visual comfort (min. 2.2%), thanks to suntunnels and skylights. Large windows facilitate the stack effect, reducing the energy demand to the HVR unit for summer cooling, while triple-glasses manage solar gains. At least, all those products are sensored and remote controlled by automatic technology and users, to adapt indoor conditions to tenants' comfort needs (Figs. 5.16 and 5.17).

Energy
Holistic efficiency is the key word to explain the project strategies. The compactness of the building minimizes heat losses, while well-insulated and air-tight prefab building components provide an adiabatic shell for extreme seasons. The HVAC system distinctly controls thermal comfort and air quality in each room, avoiding energy waste. As the energy demand is minimized, its supply is demanded to a combination of heat pump and gas furnace. A local green energy provider assures 100% renewable sources, such as wind and hydroelectricity. Thus, the building output optimal performance levels if compared to Canadian standards (Fig. 5.16).

Environment
The LCA calculation has validated the project from the production to the end-of-life phases, considering how renewable sources positively influence the environmental impact of the use-phase. GreenHouse requirements on faucets, toilet flushes and washing machines help to achieve 57% freshwater saving potential. At least, more than the 89.5% of building material has a recycling potential, and 80% of wooden products (by weight) could show SFI certifications (Fig. 5.16).

Fig. 5.15 Great Gulf AH Centennial Park (Ph. Shai Gil Fotography)

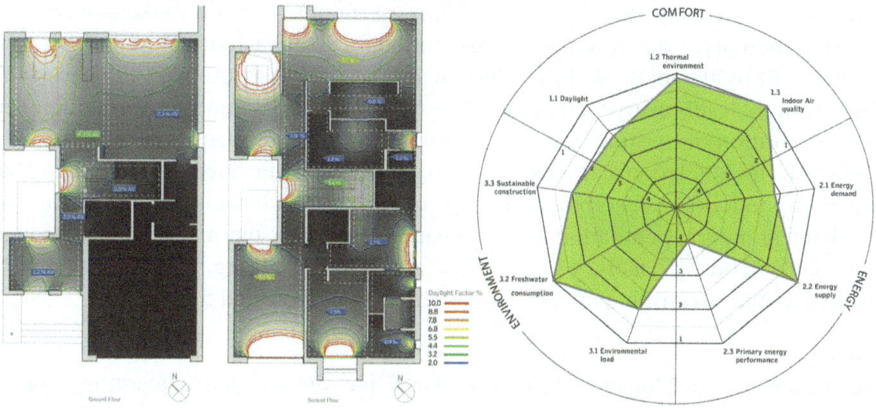

Fig. 5.16 Great Gulf AH Centennial Park DF analysis and AH Radar (© Great Gulf and AH Alliance)

Fig. 5.17 Great Gulf AH Centennial Park indoor spaces and details (Ph. Shai Gil Fotography)

Green Solution House

Location	Bornholm, Denmark
Project type	refurbishment and new construction
Use	public building—offices
Client	private
Design	**Architectural design** Kasper Guldager Jensen—3XN, Steenbergs Tegnestue, SLA, GXN innovation; **Engineering design** Ramboll, COW; **Developer** Trine Richter, Hotel Ryttegården **AH monitoring** LEAPCRAFT
Year	2015.

Green Solution House (GSH) is a conference center, whose project consists of the renovation of an existing building, through the AH vision of a new, sustainable, healthy, comfortable and creative environment (Fig. 5.18).

Comfort
In living and working spaces, natural light is essential to provide visual comfort and high productivity, reducing energy consumption for artificial light. Thus, in GSH sun-tunnels and light cables, skylights and huge fenestration assure well-lit spaces and high daylight levels everywhere (e.g. DF = 6.6% in the conference room). Moreover, the substitution of existing concrete railings with glass surfaces improves daylight conditions for hotel rooms. Besides, a green wall and specific finishing materials clean and purify indoor air, absorbing pollutant particles and balancing humidity levels. A mechanical/natural ventilation system guarantees indoor thermal comfort, through permeable acoustic panels, which diffuse air from the ceiling, with a reduction of duct sizes and performing as radiative cooling (Figs. 5.19 and 5.20).

Energy
According to the educational purpose, GSH has an interactive learning tool by LEAPCRAFT, which allows to visualize energy fluctuations and to understand the correlation between users' behavior and consequent variations of energy consumption and environmental impact. This tool has recorded data from the as-built phase to the users' interaction one, calculating an average of 3.33 kWh/m^2 as first-year energy performance. Besides, the energy demand is 2200 kWh/year, whose 32.3% comes from renewable sources. Global energy requirements are satisfied by exchanges with Bornholm local grid; floor heating/cooling are supplied on-site, by a thermal energy system partly based on food leftovers recycle. Moreover, south-located glasses are equipped with PV-cells that produce 5000 kWh/year (Fig. 5.19).

Environment
Inspired by a Cradle to Cradle life cycle concept and circular sustainability, GSH is the results of assembled solutions, whose materials and products have been inquired through an LCA perspective and certified for they recyclability (nearly 50% of total) and social responsibility of resource use. This process has been validated by DGNB.

Fig. 5.18 Green Solution House (Ph. Adam Mørk)

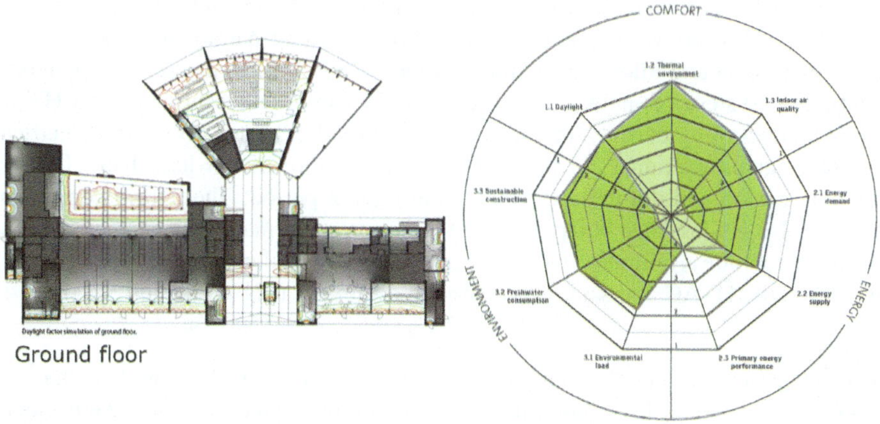

Fig. 5.19 Green Solution House FLD distribution and AH Radar, which compares the existing building and the renovation project, highlighting the optimization process of the AH parameters (© Teknologisk Institut and AH Alliance)

Fig. 5.20 GHS under construction (Ph. Laura Starner) and interior details (Ph. Adam Mørk)

OptimaHouse

Location	Kiev, Ukraine
Project type	new construction
Use	family house
Client	private
Design	**Architectural and energy design** Arch. Alexander Kucheravy; **Automatization and systems** Schneider Electric; **Structural design and construction** Dostupne Zhytlo Construction materials Metrotile, Velux, Veka, Saint-Gobain, Manezh
Year	2015.

OptimaHouse is a two-floor compact modern house, whose envelope and systems design is conceived with a holistic approach, aiming the optimal solutions in living comfort, energy efficiency, environmental impact, construction, and cost (Fig. 5.21).

Comfort

OptimaHouse reaches high scores in each AH comfort parameter. The envelope has the 31.5% of windows to total building area, providing natural light, solar gains in cold seasons and allowing obtaining a DF of 3–5% in living spaces. A hybrid ventilation system of natural cross-ventilation in middle seasons and the HVAC for heating/cooling periods, the air-air heat pump, the "water floor" and electrical ceramic panels guarantee an operative air temperature of not less than 21 °C in the cold period and not more than 24 °C in the hot period. The hybrid ventilation ensures also not less than 900 ppm of CO_2 concentration, thank sensors that measure inside temperature, humidity, CO_2 concentration, control the opening/closing of roof windows, the HVAC and sent information to occupants (Figs. 5.22 and 5.23).

Energy

The integrated design of OptimaHouse assures an energy demand less than 60 kWh/m^2 year—65% less than a traditional building in Ukraine. This achievement is due to several choices, such as the compactness of the building, its optimal orientation to the sun and insulation ($U_{wall} = 0.15$ W/m^{2}K, $U_{roof} = 0.09$ W/m^{2}K, $U_{floor} = 0.09$ W/m^{2}K, $U_{roofwind.} = 1.28$ W/m^{2}K, $U_{facade\ wind.} = 1.0$ W/m^{2}K), and the high efficient air-air heat pump. The energy demand is 45% covered by renewable sources, strictly integrated to the building: 24 vacuum solar tubes satisfy the 86% of hot water demand and the 10% for floor heating, while 6 integrated roof PV panels (2.3 m^3 polycrystalline silicon cells) produce 360 kWh/year of electric energy (Fig. 5.22).

Environment

OptimaHouse environmental impact is minimized, in terms of both consumptions than recycled potential. CO_2 emissions are less than 7 kg/m^{2}year, while freshwater consumption is reduced to 25% if compared to local legislation standards. Besides, the recycled content for all building is more than the 30%, and FSC label certifies the local origin of 80% timber products and components.

Fig. 5.21 Optimahouse (Ph. Alexander Kucheravy)

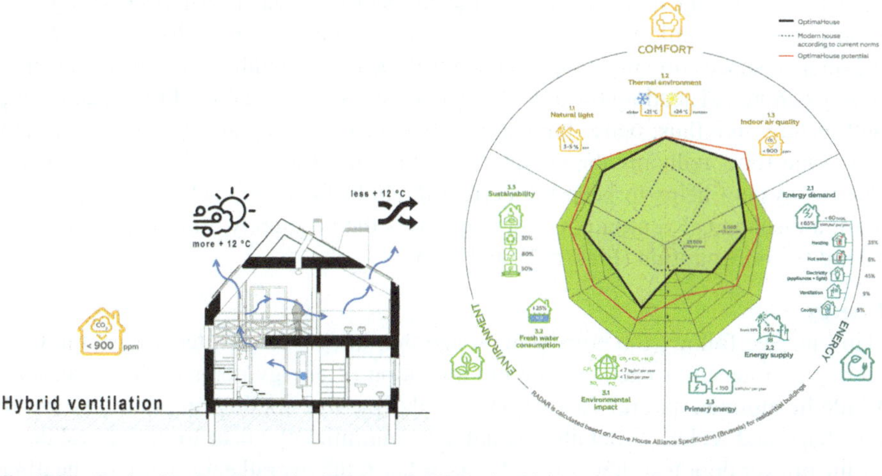

Fig. 5.22 Optimahouse schematic design and AH Radar (©Alexander Kucheravy)

Fig. 5.23 Optimahouse details: on the left, the different material treatment of the building's envelope, according to sun exposition; on the right, a detail of the 6 integrated roof PV panels (Ph. Alexander Kucheravy)

Haus am See

Location	San Felice del Benaco, Italy
Project type	new construction
Use	family house
Client	private
Design	Arch. Eileen Meyer
Year	2014.

Haus am See is a two-storey private residence in the northern Italian surrounding of the Garda Lake. The aim to best fit the building into the natural landscape lead the design strategies, which focus on the AH principles integration, beyond the passive house standards. The result is a Net Zero Energy Building and the first Italian AH, with a residential use (Fig. 5.24).

Comfort
Visual comfort is the first requirement that has been pursued, in terms of both daylight supply and panoramic view on the surrounding landscape. This disposition shapes the interior spaces organization and the envelope of the building, as the north façade is transparent and heading towards the lake. Besides, looking for a direct relationship with nature and, thus, better human well-being fits with a passive cooling strategy in summertime, reducing the energy consumption. Moreover, a domotic orientable shading system helps to keep the correct daylight factor in each indoor space. At least, a mechanical ventilation system controls the air quality, in terms of CO2 and humidity concentration (Figs. 5.25 and 5.26).

Energy
Haus am See follows the holistic approach that integrates architecture, technology, and system design, to achieve the minimization of energy demand, in all seasons. While in summer, natural ventilation provides passive cooling, in winter the building shape and technical details guarantee to minimize the heat-losses, as validated by the blower door test ($n50 = 0.22\ h^{-1}$). At least, the overall energy use for heating, cooling, hot water, ventilation and electricity is 42 kW/m^2 year (supplied by PV panels on the roof), as calculated and monitored for more than one year (Figs. 5.25 and 5.26).

Environment
As the attachment to the place of the site has come over the entire project, the designer decides for local handcrafted and traditional construction materials, as bricks and concrete. This choice leads to a significant contribution to a better indoor climate, thanks to the inertial properties of massive materials. Moreover, the proximity to the site has reduced the global emission of the construction works. In addition, wooden products used for floor and windows frames have FSC certificates (Fig. 5.25).

Fig. 5.24 Haus am See (Ph. Arch. Eileen Meyer)

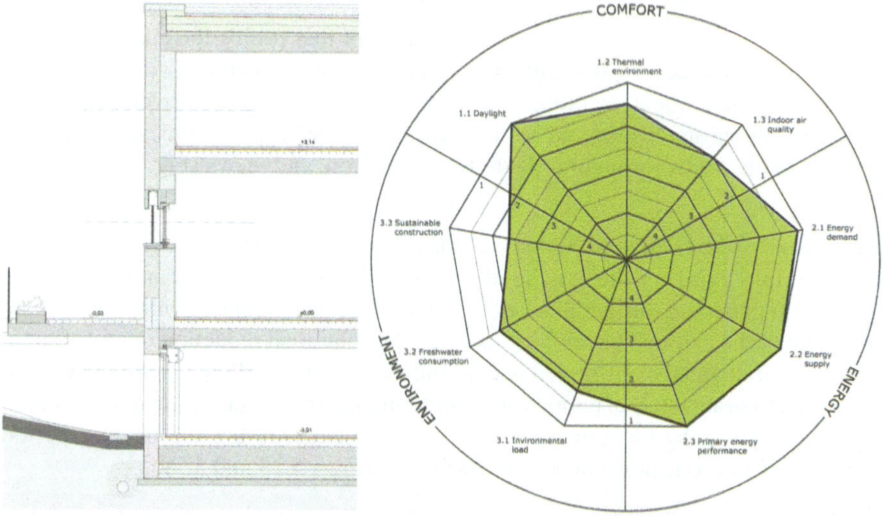

Fig. 5.25 Haus am See: schematic design and AH Radar (©Arch. Eileen Meyer and AH Alliance)

Fig. 5.26 Haus am See details: on the left, a detailed view of the indoor spaces and finishing materials; on the right, the panoramic view on Garda Lake (Ph. Arch. Eileen Meyer)

House by the Garden of Venus

Location	Willendorf, Austria
Project type	refurbishment and enlargement
Use	family house
Client	private
Design	**Planning and site supervision** Valker Dienst, Inprogress Architektur Consulting and Chistoph Feldbacher; **Static** Merz kley partner ZT GmbH; **Daylight planning** VELUX Österreich GmbH; **Artificial light planning** Podpod design; **Building concept** Active House Standard; **Wooden structure and interior construction** Kasper Greber Holz-und Wohnbau GmbH, Tischlerei Herbert Feuerstein, Tischlerei Ing. Gerhard Graschopf GmbH
Year	2013.

"House on a house" is the architectural concept that fulfills the client's needs of spatial expansion, despite the high density of the small Willendorf village. That led to a height development, an elongated tympanum that seems to overrun the existing garden on the ground floor and stretch out to the landscape. The design strategies respond to the needs for natural daylight, ventilation, and views (Fig. 5.27).

Comfort
The House by the Garden of Venus provides excellent daylight conditions, thanks to the several skylights and the large southern window, which led a large amount of direct sunlight to penetrate during all seasons. The sun screening and awning system and the shape of the external envelope guarantee the solar load reduction in midsummer, while the strategically positioned and automated windows allow a ventilate cooling strategy, that grant internal temperature values under 26 °C. A mechanical ventilation system integrate the cross natural ventilation strategy during the heating period, allowing a CO_2 concentration below 1000 ppm (Figs. 5.28 and 5.29).

Energy
The orientation of the existing building and the context constraints do not allow the installation of thermal collectors and/or photovoltaics panels strictly integrated into the building. However, the neighboring building is almost ready to host a solar system, while the heating energy comes from renewable resources, as the timber from the owners' woods. The wood furnace is going to be substituted by an efficient geothermal heat pump. The total energy consumption for heating, hot water and electricity applications is 50 kWh/m^2 year (Fig. 5.28).

Environment
The upper-storey extension results as an intelligent densification strategy, avoiding a further soil consumption. Besides, a detailed LCA certifies that the building has a low environmental impact, thanks to the water saving fittings and the employment of local wood in the lightweight construction of the new extension. This led the project to achieve the 75% of recyclable materials, at the end of building lifespan (Figs. 5.28 and 5.29).

Fig. 5.27 House by the Garden of Venus (Ph. Volker Dienst)

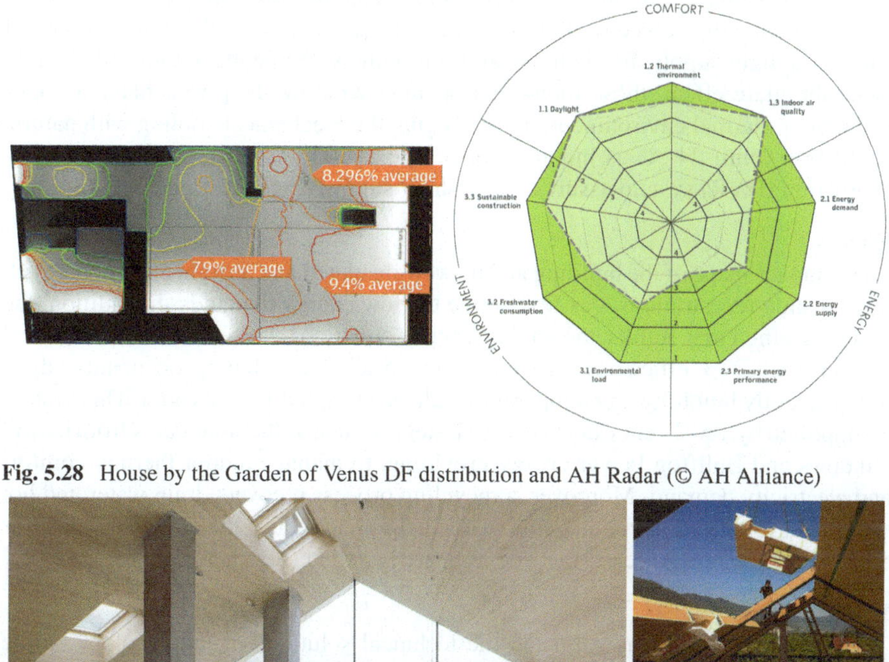

Fig. 5.28 House by the Garden of Venus DF distribution and AH Radar (© AH Alliance)

Fig. 5.29 House by the Garden of Venus: the indoor space and construction phase (Ph. Jörg Seiler)

Great Gulf Active House

Location	Toronto, Canada
Project type	new construction
Use	**family house**
Client	private company
Design	**Architectural project** Superkül; **Engineering project** Enermodal; **Developer** Great Gulf
Year	2013.

Great Gulf Active House is the result of the integration between a standard Canadian house design and the innovative AH principles, optimizing traditional architectural aspects and advanced technological devices, to provide human comfort and wellbeing, and energy efficiency while respecting the environment (Fig. 5.30).

Comfort
Great Gulf Active House architectural design welcomes the aim to provide a better indoor climate: the gabled roof allows double-height spaces, with a huge amount of natural daylight supply, by skylights and sun-tunnels. While the automated-shading avoid the glare effect, those motorized-opening windows help to achieve an optimal indoor thermal environment, interchanging the mechanical cooling with natural ventilation (until 24 °C as indoor temperature value). At least, the heat recovery ventilation (HVR) system assures the fresh air supply (Figs. 5.31 and 5.32).

Energy
The compact shape of the building and its well-insulated and airtight envelope—with low U-value fenestrations—as first define the low energy demands. In addition, the system's efficiency fulfills the saving-energy equation. A dual-zoned mechanical system, with 97% efficiency, distributes air on all floors, letting occupants adjust independently humidity and temperature values, in each different room. This system is supported by 100% renewable energy strategies, such as the solar gains from hybrid windows and Bullfrog Power energy providing, to manage indoor thermal comfort and electricity demand. Moreover, a recycling process recovers drain water and hot grey water to preheat incoming cold water, helping to reduce the energy demand for domestic hot water. At least, LED devices are installed (Fig. 5.31).

Environment
The LCA-calculation has validated the technical solutions of the envelope, while the Canadian SFI (Sustainable Forest Initiative) certifies the recyclable content of wooden products, such as the traditional frame structure. Moreover, the designers calculate that more than 50% of the construction materials would have a recycling potential. At least, a rainwater cistern could collect a large volume of water, thanks to the big roof area. This minimizes the freshwater consumption, with a saving potential of 35%, based on annual rainfall in Ontario values (Figs. 5.31 and 5.32).

Fig. 5.30 Great Gulf Active House (© Great Gulf)

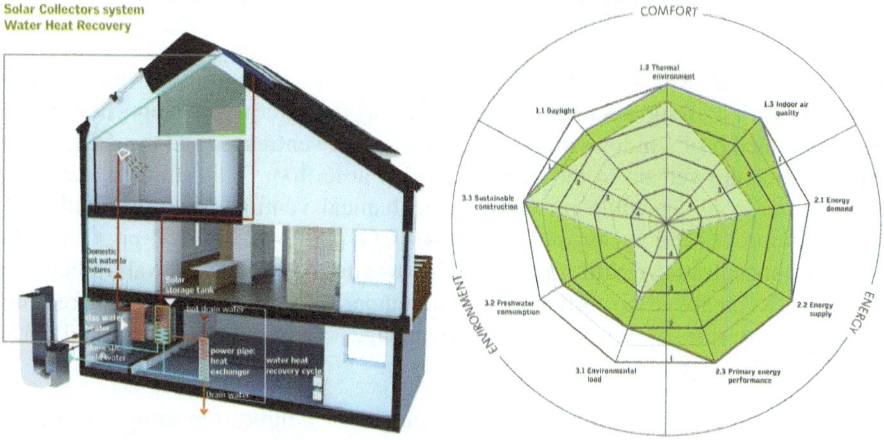

Fig. 5.31 GreatGulf AH schematic design and AH Radar (© GreatGulf and AH Alliance)

Fig. 5.32 Great Gulf Active House: the entrance and the bight indoor space (© Great Gulf)

The Poorters van Montfort

Location	Montfort, Netherlands
Project type	refurbishment and volume addiction
Use	social housing
Client	private company
Design	VELUX A/S, Danfoss; **Development partners** Bouwhulp Groep Architects, Bam Woningbouw; **Advisors** Nieman Engineering Advisors
Year	2013.

The Poorters van Montfort is a retrofit initiative held in the Netherlands, which uses the AH protocol to investigate the weaknesses of 10 existing social housings and recover them, upgrading buildings to Active Houses. It proves the balance between initial investments and final energetic, and thus economic, savings (Fig. 5.33).

Comfort
The Poorters van Montfort presents more than one solution to improve the indoor comfort. As first, designers focused on remodeling the external envelope: new façade and automated roof windows assure horizontal cross-ventilation and the stack effect, through the stairwell up to the new and widen attic floor. This strategy works for the summer cooling, while in winter a mechanical ventilation system with CO_2 sensors improves indoor air quality levels ($[CO_2]_{max} = 325 \div 350$ ppm over the outdoor concentration). Still, the large window surfaces assure DF values between 3.6 and 11%, far above the current European standards. At least, sensored and remote-controlled exterior sun-screenings help to manage solar gains (Figs. 5.34 and 5.35).

Energy
The energy approach abides by the Trias Energetica, a three-step strategy to design highly energy-efficient constructions. The first step is to reduce the energy demand, by improving the insulation levels of the envelope ($U_{fac\ window} = 0.65$ W/m^{2}K, $U_{wood\ fac} = 0.22$ W/m^{2}K, $U_{brick\ fac} = 0.27$ W/m^{2}K) and installing energy-efficient appliances (hybrid ventilation system and water-to-water heat pump). These solutions lead the building to a 57.12% reduction in the energy demand. The second and third step involves the energy supply, by on-site renewable sources, such as solar thermal collectors (4.5 m^2 on the new roof surface) or PV solar cells (19.5 m^2), or by off-site high-efficient conversation processes (e.g. cogeneration). At least, 63.9 kWh/m^2 year are supplied by 100% renewable energy, with a surplus of 2.4 kWh/m^2 year (Fig. 5.34).

Environment
Aiming an eco-friendly impact, this project follows three key-criteria: zeroing the non-renewable energy consumption, minimizing global warming and acidification potential, and the freshwater consumption. According to these purposes, an LCA analysis reveals that just the production processes of the existing building materials released 1/3 of CO_2 and 2/3 of SO_2 and NO_X emissions (Fig. 5.34).

Fig. 5.33 The Poorters van Montfort (Ph. Adam Mørk)

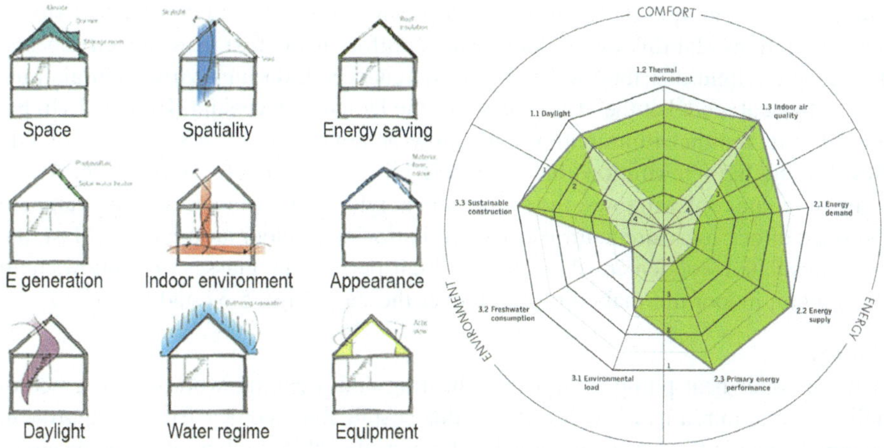

Fig. 5.34 The Poorters van Montfort schematic design and AH Radar (© VELUX Group and AH Alliance)

Fig. 5.35 The Poorters van Montfort: the new attic space (Ph. Adam Mørk)

Haus am Moor

Location	Krumabach, Austria
Project type	new construction
Use	private residence and studio
Client	private
Design	**Architectural design** Berardo Bader Architects; **Engineering design** Gerhard Bilgeri; **Daylight analysis** Donau-Universität Krems and VELUX Group
Year	2012.

A unique wooden volume stretches through the moor, between countryside and village, to host a living and working space under the same roof. Regional architectural elements and natural, untreated materials are combined into the design, aiming high comfort, energy efficiency and respect for the environment (Fig. 5.36).

Comfort
The need for a unique building to host a private residence and an attached studio puts daylight availability and indoor visual comfort as the first investigated aspects. Thus, the daylight distribution has been evaluated both through virtual simulations and a physical model at the "Lichtlabor" of the Donau-Universität Krems. Skylights provide the zenithal illumination and allow achieving $3.5 \div 4.9\%$ levels of DF. Besides, the big fenestrations on the façade bring users directly into nature. At the same time, the roof windows, the shape, and height of the wooden ceiling guarantee the stack effect, to assure natural cross ventilation and indoor air quality. At last, the hyper-insulated wooden envelope and the high-performant glasses of windows help to maintain indoor thermal comfort, all over the year (Figs. 5.37 and 5.38).

Energy
A brine water heat pump with a floor heating and a central wood-burning hearth deliver the required heat. According to this sustainable and eco-friendly approach, designers decided to use different natural materials: while timber envelope provides a good insulation level, the massive cement-based walls stock heat, during cold seasons. Besides, a mechanical ventilation system recovers heat in winter, to minimize energy waste; it turns off in hotter seasons when the opening of windows activates the natural ventilation, to allow a passive cooling and discharge massive walls from thermal overloads. The total energy requirement is 25 kWh/m^2 year (Fig. 5.37).

Environment
The integration with the natural surrounding is the key point of this project. Materials, construction process, products have been chosen because of that. The envelope is furnished with untreated timber, derived from local pines. The floor slabs are produced from locally unearthed clay, which is dried on-site and pressed to form special-shaped bricks that host the heat pipes. Using local materials helps to reduce "grey energy" and CO_2 emissions, due to a long way transportation (Fig. 5.37).

Fig. 5.36 House on the Moor (Ph. Jörg Seiler)

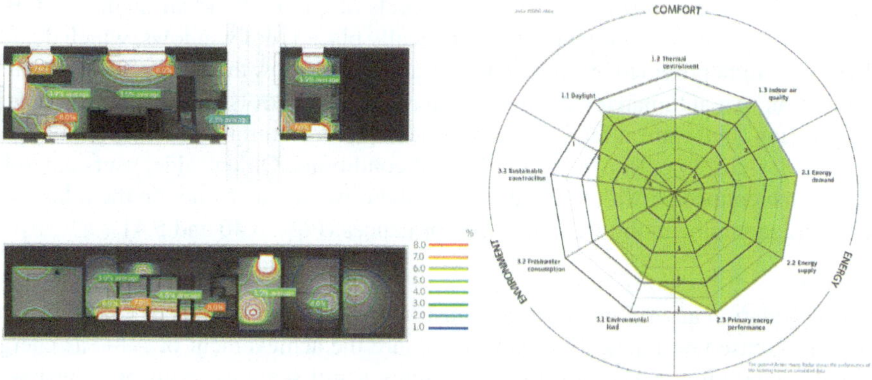

Fig. 5.37 House on the Moor DF distribution and AH Radar (© Jörg Seiler and VELUX Group)

Fig. 5.38 House on the Moor in visually connected with the outdoor environment, gaining direct and diffuse daylight (Ph. Adolf Bereuter)

ISOBO Aktiv

Location Sandned, Norway
Project type new construction
Use family house
Client private company
Design **Architectural concept** SFAS|arkitektur **Project management** Jadarhus AS by Kurt Hobberstad **Partners** VELUX Norge AS
Year 2011.

ISOBO Aktiv is an innovative design approach to develop and build low-energy houses. It uses demo-buildings to test different technical solutions on full-scale, and define the measures that more influence indoor climate, energy and environment (Fig. 5.39).

Comfort
Optimal indoor comfort depends by high levels of daylight and air quality. ISOBO Aktiv demo-house is equipped with strategically placed roof windows, which double the natural light provided by vertical fenestrations and allow the stack effect of natural ventilation. An automated steering system remotely controls sunscreens, to prevent overheating and glare effect, windows opening, for natural cross ventilation, and heating devices, to optimize indoor thermal conditions. On one side, users could be every time connected with the building, by their own smartphone; on the other one, the house could be monitored, in its performances (Figs. 5.40 and 5.41).

Energy
Extra insulation and tightness of the construction envelope, and active strategies on systems—based on renewable sources—assure the achievement of A++, as energy rating for buildings. The envelope represents a full-scale experiment to test several possible technical solutions, with U-values between 0.10 kWh/m^2K (of super-insulated external walls and sloped roof) and 1.00 Wh/m^2K of roof windows. Thus, the performances about heat leakages have been tested on the finished house, recording a value of N50 = 0.35/h. Besides, 8 m^2 solar collectors, 8 solar cell panels, an air-to-water heat pump, with heat recovery, supply almost 90% of required energy (44.4 kWh/m^2 year). A ground collector under the house helps the ventilating unit, increasing its efficiency and naturally heating/cooling the conditioned air (Fig. 5.40).

Environment
The house is based on timber-frame construction system, industrially produced and prefabricated. All construction units were delivered as pre-cut and ready-for-assembly. This choice helped to reduce construction waste because all the components have been specifically designed, through an optimized process. Local partners and suppliers minimized long-distance transportation, and consequent CO_2 emission during the construction phase (Fig. 5.40).

Fig. 5.39 Isobo Aktiv (© Jadarhus AS)

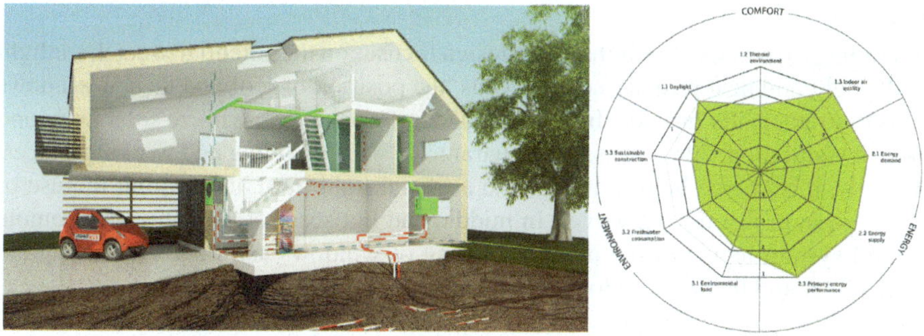

Fig. 5.40 Isobo Aktiv schematic design and AH Radar (© Jadarhus AS and AH Alliance)

Fig. 5.41 Isobo Aktiv details: on the left, the indoor bright space, thanks to huge roof windows; on the right, the solar panels and collectors on the roof, which provide almost the 60% of required enrgy (© Jadarhus AS)

Sunlighthouse

Location	Pressbaum, Austria
Project type	new construction
Use	family house
Client	private company
Design	**Architectural concept** HEIN-TROY Architekten; **Energy conceptstructural calculations and ecological evaluations** Donau-Universität Krems and Institute for Healthy and Ecological Buildings; **Building development** VELUX Group
Year	2010.

Sunlighthouse is the first Austrian carbon-neutral single-family house, thus fulfilling Model Home 2020 aims. Its iconic sloping roof combines interesting space definition with high levels of indoor comfort and energy efficiency, taking advantage of the sun for daylight and renewable energy supply (Fig. 5.42).

Comfort
According to its name, Sunlighthouse design focuses on the potential to catch sunlight as daylight and heat primary source. Thus, the position of roof and façade windows aims to strategically provide the best visual comfort and to maximize solar gains and natural ventilation. 72 m^2 glass area guarantee 5% DF in living and working spaces. Actually, this choice help to reduce the artificial light need and to limit the use of HVAC system just during winter. In middle and hot seasons, an automated remote control system regulates the opening/closing of windows and blinds, fostering the stack effect and the indoor global comfort (Figs. 5.43 and 5.44).

Energy
Sunlighthouse annual energy demand is 50.8 kWh/m^2 year, thanks to technical solutions and high efficient products ($U_{w,roof}$ = 1.1 W/m^2K, $U_{w,façade}$ = 0.76 W/m^2K, U_{wall} = 0.13 W/m^2K, U_{roof} = 0.12 W/m^2K). This requirement is satisfied just by renewable energy sources: a highly efficient brine/water heat pump, thermal solar collectors, PV solar cells system and highly energy-efficient appliances guarantee an annual energy surplus of 12.2 kWh/m^2 year. This means that until 30 years after construction, the house will be a real carbon neutral building. The required energy will be balanced by the clean energy produced by systems (Fig. 5.43).

Environment
Sunlighthouse perfectly fits its surroundings. The building outdoor and indoor shapes, its finishing and overtures on the envelope testify the deep connection with the Vienna woods and mountains environment. Moreover, in order to prove the positive-energy and eco-friendly design (CO_2-neutrality and ecological materials), Sunlighthouse has been monitored in its use-phase by VELUX Austria, Danube University Krems and the Austrian Institute for Healthy and Ecological Building (Fig. 5.43).

Fig. 5.42 Sunlighthouse (Ph. Adam Mørk)

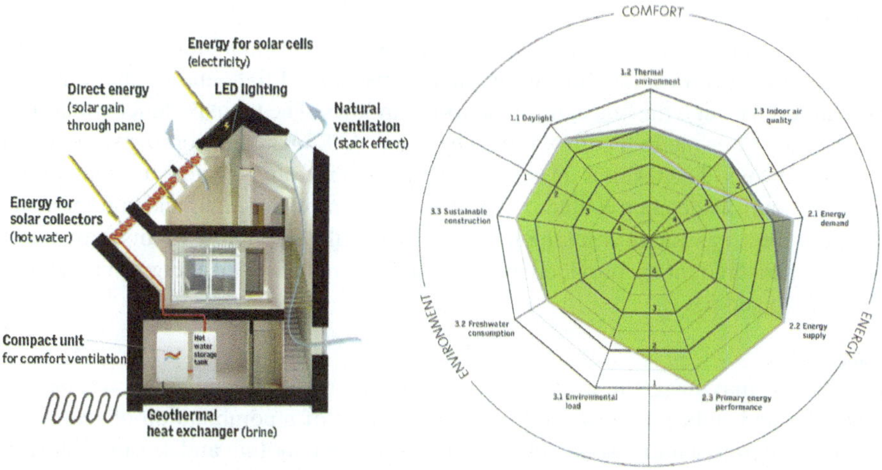

Fig. 5.43 Sunlighthouse schematic design and AH Radar (© VELUX Group and AH Alliance)

Fig. 5.44 Sunlighthouse: the bright indoor spaces and the outdoor peculiar shape (Ph. Adam Mørk)

Green Lighthouse

Location København, Denmark
Project type new construction
Use office building
Client private company
Design **Architectural project** Christensen & Co Architects; **Engineering project** Cowi **Developer** Danish University and Property Agency
Year 2009.

As a showpiece for the UNI Climate Change Conference of Copenhagen, 2009, Green Lighthouse is the first Danish public CO_2-neutral building, which is hosting the Faculty of Science of the University of Copenhagen. It belongs to the six demo-houses of the Model Home 2020 project, whose aim is the optimal balance between architectural quality, healthy indoor climate, and energy efficiency (Fig. 5.45).

Comfort
Green Lighthouse conveys combined architectonic and sustainable solutions. Thanks to its circular atrium space, it hosts an internal passage of light and lets natural cross ventilation, through the house from the roof windows. The main parameter that rules the design of this internal core was the DF, at least 3% in all the working stations and 2% in hallways. The result is an active energy saving strategy, where lux sensors and dimmer controls detect the inner level of naturally provided light and communicate with the electric system to set the power of artificial LED lights. In this way, visual comfort and daylight are provided, with a four times reduction of energy consumption (Figs. 5.46 and 5.47).

Energy
In order to maximize the energy saving, Green Lighthouse integrate architecture materials and light into an overall plan. It consists of a combination of renewable sources: district heating, solar cells, solar heating and cooling and seasonal storage. Solar collectors on the roof and the geothermal heat system in the ground convey 35% of energy via heat pump, while 65% of it comes from the eco-friendly district heating, which shares the 35% of renewable energy. At least, 76 m^2 PV cells guarantee electricity for lighting, hybrid ventilation, and pumps. This energy concept leads to 75% energy consumption cut-off (30 $kW/m^2/year$), and thus it represents a model for office buildings, to become CO_2 neutral construction of the future (Figs. 5.46 and 5.47).

Environment
The path of the sun and its movement around the house inspires the architectural concept. This choice emphasizes how sunlight is considered as the most significant energy sources of the building, driving the definition of system devices and, finally, leading to a neutralization of CO_2 emission (Fig. 5.46).

Fig. 5.45 Green Lighthouse (Ph. Adam Mørk)

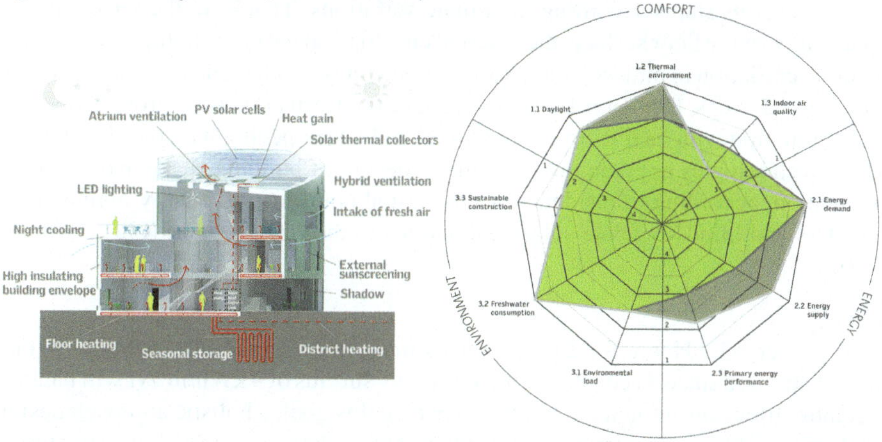

Fig. 5.46 Green Lighthouse schematic design and AH Radar (© VELUX Group, AH Alliance)

Fig. 5.47 Green Lighthouse details: the bright and circular atrium space and the solar devices on the roof (Ph. Adam Mørk)

Home for Life

Location	Aarhus, Denmark
Project type	new construction
Use	family house
Client	private company
Design	**Architectural project** AART A/S; **Engineering project** Esbensen Rådgivende Ingeniører; **Executive developer** VELFAC & VELUX Group
Year	2009.

Home for Life belongs to the six experimental demo-houses of the Model Home 2020 project by VELUX Group. It aims to demonstrate the feasibility of climate neutral constructions, with a high level of livability (Fig. 5.48).

Comfort
An automated system records data from heat, CO_2 and humidity sensors, to control the active envelope and systems. The slate-covered façade is a dynamic screen that changes its status according to climate variations. Thanks to the wide window area—40% of the floor surface, the external envelope opens to daylight, solar gain and natural ventilation in mid and hot seasons, or becomes an adiabatic layer, against winter losses. Besides, the control system manages the fresh air needs, through mechanical ventilation with heat recovery in winter, and auto-opening windows in summer. These technologies upgrade the bioclimatic design of building shape and windows orientation, according to residents' needs, visual comfort, and energy optimization (shutters and blinds regulate solar heat and avoid the risk of glare) (Figs. 5.49 and 5.50).

Energy
As a proper AH, Home for Life produces more energy than it consumes. Within almost 40 years since its construction, the energy surplus of 9 kWh/m^2/yr will balance the entire lifecycle amount of energy. To pursue this goal, a holistic approach design qualitative (shape, orientations) and quantitative (dimensions, technical features) parameters (Fig. 5.49). Remote controlled fenestrations and shading system optimize light, air and heat gain, achieving about 50% of heating requirements by passive solar heat in winter. Natural and mechanical ventilation ensure fresh air in any season. 50 m^2 solar cells, 7 m^2 solar collectors, and a heat pump produce electrical energy, hot water, and space heating. At least, the control system optimizes energy consumption, scheduling systems use just when they are needed.

Environment
Durable and low emissions materials and renewable energy-based systems help to reduce the environmental impact of the house, while the measured data improve the simulated performances (Fig. 5.49). Actually, the combination of natural and advanced technological solutions makes it be a CO_2 neutral building.

Fig. 5.48 Home for Life (Ph. Adam Mørk)

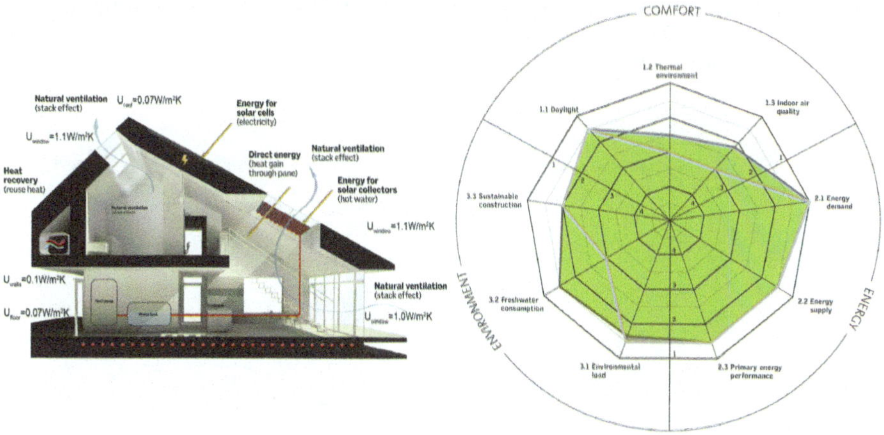

Fig. 5.49 Home for Life schematic design and AH Radar (© VELUX Group and AH Alliance)

Fig. 5.50 Home for Life details: on the left, the bright interior spaces, where the wide windows let the view of the surrounding landscape; on the right, the active façade, covered by black slates, which integrate the dark surfaces of the PV cells and solar panels (Ph. Adam Mørk)

Bibliography

The Directive 2010/31/EU of the European Parliament and of the Council of 19 May 2010 on the energy performance of buildings (2010) Official Journal of the European Union 53

Marszal AJ et al (2011) Zero energy building—a review of definitions and calculation methodologies. Energy Build 43:971–979

Bell J (2004) A mixed-use renaissance. Mortgage Banking 5(8):66

Flynn L (2003) Developing a winning combination. Build Des Constr 44(3):26–32

BPIE (2015) Nearly zero energy buildings definitions across Europe, 2015

Sesana MM, Salvalai G (2013) Overview on life cycle methodologies and economic feasibility for nZEBs. Build Environ 67:211–216

Active House standard. Available at: https://www.activehouse.info/

Active House Alliance (2017) Active house award. https://www.activehouse.info/. Accessed Jan 2018

Eriksen KE (2016) Active house—a tool for development of CO_2 neutral and sustainable buildings. PLEA Los Angeles 07:2016

Eriksen KE, Rode C (2013) Active house specification—evaluation of comfort, energy and environment in buildings REHVA J 03

https://www.activehouse.info/

The manufacturer's authorised representative in the EU is Springer Nature Customer Service Centre GmbH, Europaplatz 3, 69115 Heidelberg, Germany. If you have any concerns regarding our products, please contact ProductSafety@springernature.com

Printed and bound by CPI Group (UK) Ltd, Croydon, CR0 4YY
30/09/2026
02225261-0013